AF552338

Springer Series in
Computational
Mathematics

22

Pshenichnyi, B. N.

B. N. Pshenichnyj

The Linearization Method for Constrained Optimization

Translated from the Russian by Stephen S. Wilson

With 6 tables

Springer-Verlag
Berlin Heidelberg New York
London Paris Tokyo
Hong Kong Barcelona
Budapest

Boris N. Pshenichnyj
Institute of Cybernetics of the
Academy of the Ukrainian SSR
142/144-4 Oletiya Oktyabrya ul.
25227 Kiev - 207
Ukraina, CIS

Translator
Stephen S. Wilson
Fourwinds
79 New Barn Lane
Prestbury
Cheltenham
Gloucestershire GL 52 3LE
United Kingdom

Mathematics Subject Classification (1991):49Kxx, 49M35, 49M40, 49M45, 65K05, 65K10, 90C05, 90C20, 90C25

Nauka, Moscow 1983: B. N. Pshenichnyj, Metod Linearizatsii

ISBN 3-540-57037-3 Springer-Verlag Berlin Heidelberg New York
ISBN 0-387-57037-3 Springer-Verlag New York Berlin Heidelberg

Library of Congress Cataloging-in-Publication Data

Pshenichnyĭ, B. N. (Boris Nikolaevich)
[Metod linearizatsii. English]
The linearization method for constrained optimization/B. N. Pshenichnyj; translated from the Russian by Stephen S. Wilson.
p. cm. – (Springer series in computational mathematics; 22)
Includes bibliographical references and index.
ISBN 3-540-57037-3. – ISBN 0-387-57037-3
1. Nonlinear programming. I. Title. II. Series.
T57.8.P7913 1994 519.7'6–dc20 94-1938 CIP

Printed in Germany

41/3140 – 5 4 3 2 1 0 – Printed on acid-free paper

Foreword

It is customary in a foreword to write about the practical importance of the problem and to describe the contents of the book. However, the practical importance of solving optimization problems has been unquestioned for some time. A wealth of scientific and popular scientific literature is devoted to this subject, so that it is almost unnecessary to repeat yet again that the theory and techniques of optimization are applied in many problems in economics, automatic control, engineering etc. On the other hand, a comprehensive summary of the contents of the book is given in the first introductory section. Therefore, I shall not dwell on this here. Instead, I shall allow myself a number of general remarks on the theory and the numerical techniques of optimization and the interlinking of these in the solution of a complicated real problem.

Electronic computers were first applied to solve optimization problems in their earliest days. The first such applications involved linear programming problems with a very simple structure, which could be solved by regular methods, and comparatively uncomplicated nonlinear problems, which were solved based on simple inductive considerations and using the facility for vast numbers of numerical computations provided by the computer technology. But the challenge of more and more new problems of increasing size and nonlinear complexity meant that it was no longer possible to use simple techniques and raw power alone. The solution of substantial nonlinear problems required a deeper theoretical study of methods for solving them together with delicate and complicated techniques for obtaining numerical results for each specific problem within a reasonable time. The process of generating such an arsenal of techniques for solving optimization problems has been intently pursued over the last twenty years.

The linearization method, which is the subject of this book, is one of the many fruits of this process.

In fact, as will be seen from what follows, the linearization method is closely related to Newton's method for solving systems of linear equations, to penalty function methods and in particular, to methods involving nonsmooth penalty functions, and, in connection with the latter, to methods of nondifferentiable optimization. The linearization method cannot be successfully applied without the efficient solution of quadratic programming problems; this leads to a connection with conjugate gradient methods and variable metrics. The desire to solve large-scale problems requires the application of techniques developed in linear programming, i.e. techniques for working with sparse matrices, a multiplicative representation of inverse matrices etc.

Finally, it is impossible to study areas and rates of convergence without applying the theory of necessary conditions for extrema, Lagrange multipliers and functions, and the concept of the dual problem. It follows that a comprehensive study of the properties of the linearization method is impossible without applying a wide range of concepts ranging from abstract theory to specifics of the computer-based implementation. Naturally, all this is reflected, to a greater or lesser extent, in the book. These questions cannot not all receive the same attention if the book is to be held to a reasonable size. However, while some problems are only considered superficially (for example, conjugate gradient and variable metric methods, methods for working with sparse matrices), this does not mean that they are unimportant. The contrary is often true, as the example of sparse matrices shows. Were it not for the use of sparse matrices, the time taken for large problems would increase catastrophically.

There is yet another very important reason why specialists who are primarily interested in a specific application of the method should nevertheless be familiar with the associated concepts and ideas. The fact is that the method is based on the solution of a general nonlinear programming problem and thus considerable time may be expended on solving subproblems of this problem. A specific problem (or class of problems) always has specific properties (for example, most constraints have a very simple form) and taking these properties into account by changing certain modules of the algorithm may lead to a considerable reduction in the time taken and the computer storage used. Clearly, such changes cannot be successfully implemented without an understanding of the principal mechanisms which lead to convergence of the algorithm.

Although this book was dictated by an urge to undertake some form of research relating to the linearization method, the author is confident that it will not be the last word on this topic. This is confirmed by the ever increasing number of papers on this and related subjects. Moreover, there are many problems which require deeper and more detailed study. In particular, these include the choice of a rule for transition from the simple to the accelerated linearization method, more precise definition of rules for choosing the step in such a transition, algorithms to change the constants in penalty functions, and techniques for approximating the matrix of second derivatives of Lagrange functions. More detailed consideration should be given to the specific properties of the linearization method when applied to large-scale problems, methods for decomposing the original and the auxiliary problems etc. All these problems are nontrivial, and the author will be pleased if this book serves not only to expand the sphere of practical application of this method but also as a starting point for further improvements and developments.

B.N. Pshenichnyj

Contents

1. Convex and Quadratic Programming 1
1.1 Introduction ... 1
1.1.1 The Linearization Algorithm ... 1
1.1.2 Convergence of the Algorithm ... 3
1.1.3 General Remarks ... 4
1.1.4 Notation ... 5
1.2 Necessary Conditions for a Minimum and Duality ... 6
1.2.1 Convex Sets ... 7
1.2.2 Convex Functions ... 7
1.2.3 Foundations of Convex Programming ... 10
1.2.4 Duality in Convex Programming ... 15
1.2.5 Necessary Conditions for Extrema. General Problem ... 16
1.2.6 Necessary Conditions for Extrema: Second Order Conditions ... 18
1.2.7 Minimax Problems ... 18
1.2.8 Penalty Functions ... 19
1.3 Quadratic Programming Problems ... 22
1.3.1 Conjugate Gradient Method ... 23
1.3.2 Conjugate Gradient Algorithm ... 25
1.3.3 Existence of a Solution ... 25
1.3.4 Necessary Conditions for an Extremum and the Dual Problem ... 28
1.3.5 Application, Projection onto a Subspace ... 30
1.3.6 Algorithm for the Quadratic Programming Problem ... 31
1.3.7 Computational Aspects ... 37
1.3.8 Algorithms for Simple Constraints. Generalization ... 41

2. The Linearization Method 43
2.1 The General Algorithm ... 43
2.1.1 Main Assumptions ... 44
2.1.2 Formulation of the Algorithm ... 44
2.1.3 Convergence of the Algorithm ... 45
2.1.4 Computational Aspects ... 49
2.1.5 Generalizations ... 51
2.1.6 The Linear Programming Problem ... 53
2.1.7 The Linearization Method with Equality-Type Constraints ... 57

2.1.8 Simple Constraints 58
2.1.9 Choice of Parameters in the Linearization Method. Modified Algorithm 60
2.2 Resolution of Systems of Equations and Inequalities 65
2.2.1 The Auxiliary Problem 66
2.2.2 The Algorithm 66
2.2.3 Convergence of the Algorithm 67
2.3 Acceleration of the Convergence of the Linearization Method 75
2.3.1 Main Assumptions 75
2.3.2 Local Analysis of the Auxiliary Problem 76
2.3.3 Preliminary Lemmas 81
2.3.4 The Linearization Algorithm and Acceleration of Convergence 84
2.3.5 Linear Transformations of the Problem 87
2.3.6 Modifications of the Linearization Method 90

3. The Discrete Minimax Problem and Algorithms 99
3.1 The Discrete Minimax Problem 99
3.1.1 The Auxiliary Problem 100
3.1.2 Some Bounds 102
3.1.3 Algorithms 104
3.1.4 Algorithm for $A_k = I_n$ 106
3.1.5 Acceleration of Convergence in the Convex Case 111
3.2 The Dual Algorithm for Convex Programming Problems 114
3.2.1 The Dual Algorithm 115
3.2.2 Bounds on the Rate of Convergence 118
3.2.3 An Algorithm for Convex Programming Problems 123
3.3 Algorithms and Examples 129
3.3.1 The Linearization Method 129
3.3.2 The Accelerated Linearization Method 132
3.3.3 Examples of Calculations 134

Appendix: Comments on the Literature 143

References 145

1. Convex and Quadratic Programming

1.1 Introduction

With the exception of Section 3.2, this book is entirely devoted to a single method of solving nonlinear programming problems, namely the linearization method. In this, it differs from most books on this subject, which usually consider various methods. The various algorithms and approaches described in the literature are not random. Many years of experience of solving nonlinear optimization problems have led specialists to the almost unanimous view that it is impossible to develop a universal algorithm, which could be uniformly and successfully applied to solve all problems. The author fully agrees with this view. Indeed, nonlinear optimization problems are extremely varied. They differ in their nonlinear structure, in the number of variables and constraints and in the amount of storage required. Practical experience shows that there exist classes of problems for which, thanks to a simple implementation and the specific nature of individual problems, a method which is apparently very ineffective from a theoretical point of view may give good results. In this respect, overall, practical computations require a number of different algorithms. The concentration of this book on a single technique reflects a desire to explain its properties and capabilities in detail and to identify its practical characteristics and advantages. The increasing practical use of the linearization method over many years has demonstrated that it is highly effective for solving very broad classes of problems. Thus, here, we shall attempt to identify its most characteristic features which provide for its high effectiveness.

1.1.1 The Linearization Algorithm

A more precise statement of the problem and the conditions on it will be given later. For the moment, we shall not restrict ourselves by particular mathematical rigour.

Thus, suppose that $I = \{1, \ldots, m\}$ is a finite set of indices. We shall consider the problem of finding a minimum of the function $f_0(x)$ subject to the constraints $f_i(x) \leq 0$, $i = 1, \ldots, m$. More concisely,

$$\min_x \{f_0(x) | f_i(x) \leq 0, \; i = 1, \ldots, m\}. \tag{1.1}$$

Although it is not particularly difficult to consider constraints of the form $f_i(x) = 0$ within the linearization method, for simplicity, we shall not do so here.

By analogy with the usual Newton method for solving systems of nonlinear equations, we attempt to linearize the nonlinear problem (1.1) at the point x and to compute an increment in the argument by solving the corresponding linear problem:

$$\min_p\{f_0(x) + (f_0'(x), p) | f_i(x) + (f_i'(x), p) \leq 0, \ i = 1, \ldots, m\}, \tag{1.2}$$

where $f_i'(x)$ denotes the gradient of the function $f_i(x)$. For the new approximation, we may take the point $x + p$. However, as a rule, (1.2) will not have a solution, and its minimum will be equal to $-\infty$. This is because a linear approximation to a nonlinear function is only valid in some neighbourhood of the point x. Thus, it is natural to try to take this fact into account. This may be done in two ways: firstly by direct bounding of the norm of the vector p, i.e. by supplementing (1.2) with the constraint

$$\|p\| \leq \delta, \ \ \delta > 0 \tag{1.3}$$

or secondly by penalizing large deviations. The second technique leads to the following auxiliary problem:

$$\min_p\{(f_0'(x), p) + 1/2\|p\|^2 | f_i(x) + (f_i'(x), p) \leq 0, \ \ i = 1, \ldots, m\}. \tag{1.4}$$

In what follows we shall consider only the second technique.

Methods based on the introduction of a constraint (1.3) in the auxiliary problem are often discussed in the literature; in particular, these are studied in detail in [37].

Thus, if x is any approximation to the solution of problem (1.1) and $p(x)$ is a solution of problem (1.4), we look for the next approximation in the form $x+p(x)$. However, it is well known that, even when solving systems of equations by Newton's method, choosing the next approximation in this way guarantees at best only local convergence, i.e. convergence from a sufficiently good initial approximation. In practice, it would be desirable to guarantee convergence from a wide region of initial approximations, and best of all, from an arbitrary initial approximation. Thus, we shall look for the next approximation in the form $x_1 = x + \alpha p(x)$, where α is a number in the half-interval $(0, 1]$, which determines a step in the direction $p(x)$. Naturally, in general, α will depend on x and there arises the new problem of selecting this step correctly.

In unconstrained optimization problems, correct choice of the step is determined by the condition that the function to be minimized should decrease substantially in the given direction. However, in the present case, we have a minimization problem with constraints and we cannot simply take the function to be minimized for this purpose. Instead, we take the goal function of the problem, plus a penalty for violating the constraints. More precisely, we suppose that

$$
\begin{aligned}
F(x) &= \max\{0, f_1(x), f_2(x), \ldots, f_m(x)\}, \\
\Phi_N(x) &= f_0(x) + NF(x).
\end{aligned}
$$

We shall choose the step α as follows. We set $\alpha = 1$ and divide this value by two until the inequality

$$\Phi_N(x + \alpha p(x)) \leq \Phi_N(x) - \epsilon\alpha\|p(x)\|^2, \quad 0 < \epsilon < 1 \tag{1.5}$$

is satisfied. Then, the first value of α for which inequality (1.5) is satisfied is taken to be the step.

It turns out that if

$$N \geq \sum_{i=1}^{m} u^i(x), \tag{1.6}$$

where the $u^i(x)$ are the Lagrange multipliers of the auxiliary problem (1.4), then the inequality (1.5) will be satisfied after a finite number of divisions.

Thus, the linearization algorithm is the following. Given the approximation x_k, the next approximation may be constructed by solving problem (1.4) for $x = x_k$ and setting $p_k = p(x_k)$ and

$$x_{k+1} = x_k + \alpha_k p_k$$

where α_k is chosen to satisfy the inequality (1.5) with $x = x_k$, using the method described above.

1.1.2 Convergence of the Algorithm

Next we consider the convergence of this algorithm. As shown in Section 2.1, satisfaction of the inequality (1.6) is crucial for convergence of the algorithm, i.e. a correct choice of N is crucial. Of course, N is not known a priori. However, problem (1.4) does not involve the number N. Thus, the values $u^i(x)$ may be calculated independently of N. By checking that inequality (1.6) is satisfied, we may correct this number and increase it if the initial choice is unsuccessful.

If the number N is suitably corrected, the algorithm converges in the following sense. The sequence of points x_k generated by it has limit points which satisfy all the constraints of problem (1.1) and the necessary conditions for a minimum are satisfied at these points. In particular, if we consider a convex programming problem, any such limit point is a solution. Moreover, if we consider a linear programming problem, it can be shown that the algorithm converges after a finite number of steps.

Next, we note that the algorithm may be applied to solve systems of inequalities. Indeed, if we assume that $f_0(x) = 0$, then solution of problem (1.1) is equivalent to simply solving the system of inequalities $f_i(x) \leq 0$, $i = 1, \ldots, m$. As shown in Section 2.2, with a small adjustment to the choice of algorithm step, under reasonably natural assumptions, in this case, the convergence will be quadratic, i.e. like the usual Newtonian method for solving systems of equations. We note that problem (1.1) also has the same property when only n independent constraints $f_i(x) \leq 0$ are satisfied as equalities, where n is the dimension of the vector x.

In Section 3.1, we consider the discrete minimax problem, i.e. the problem of minimizing the function

$$F(x) = \max_{1 \leq i \leq m} f_i(x)$$

and give a modification of the linearization algorithm suitable for solving this problem. In this case, it turns out that the convergence of the algorithm is quadratic, if the minimum of the function $F(x)$ is a so-called Chebyshev point; this assumption is usually satisfied in problems of best uniform approximation. In passing, we note that, experience shows that, even without this assumption, the algorithm is highly effective when applied to minimax problems.

In the general case, the linearization algorithm only converges with the rate of a geometric progression. It is quite easy to see this from the fact that for an unconstrained problem it turns into the usual gradient descent method, for which such convergence is rigorously proven. In some problems, convergence with the rate of a geometric progression may be unsatisfactory. Thus, in Section 2.3, we consider a number of ways of accelerating the convergence. We mention just one of these ways here. In this case, at each iteration step, instead of (1.4), we solve the problem

$$\min_p \{(f_0'(x_k), p) + 1/2(A_k p_k, p_k) | f_i(x_k) + (f_i'(x_k), p) \leq 0, \quad i = 1, \ldots, m\}.$$

In Section 2.3, the matrix A_k is chosen to be the matrix of the second derivatives with respect to x of the Lagrange function

$$L(x, u) = f_0(x) + \sum_{i=1}^{m} u^i f_i(x)$$

and it is shown that, under specific assumptions, this leads to convergence at a rate faster than an arbitrary geometric progression. But the computation (even using difference formulae) of the matrix of second derivatives may be very laborious. Thus, it would be interesting to carry out further studies of methods for recursive transformation of the matrix A_k into A_{k+1}, as, for example, in [46]. In this case, in order to guarantee accelerated convergence, it is sufficient to guarantee that the A_k converge to the matrix of second derivatives of the Lagrange function at the solution.

1.1.3 General Remarks

Next, we turn our attention to a number of general points associated with the linearization method. It follows from the above that this method combines the features of a number of known methods, including Newton's method, the method of penalization functions and the standard gradient descent method. In particular, the linearization method may be viewed as a method of minimizing a penalty function $\Phi_N(x)$. It is known (see Section 1.2) that, for sufficiently large N, the function $\Phi_N(x)$ has the property that, under certain assumptions, its minima coincide with the solutions of problem (1.1). However, this function is not differentiable and the usual effective methods of unconstrained optimization do not apply to it. The linearization method may be used to locate its minima; here, the direction of the displacement p between points is obtained by solving problem (1.4) and is independent of the choice of N. Moreover, the choice of this direction is invariant with respect to scaling of the constraints $f_i(x)$, $i = 1, \ldots, m$, i.e. with respect to replacement of the functions $f_i(x)$ by functions $a_i f_i(x)$, $a_i > 0$; this is important as far as computation is concerned.

In studying the method, we need to use a whole arsenal of facts from optimization theory. Thus, Section 1.2 contains basic theorems from the theory of convex functions and gives necessary conditions for extrema; it may serve as a handbook and may be omitted by the well-informed reader. While Section 1.2 includes many theorems without proofs, detailed proofs are given in all other sections. This is not simply out of a desire to maintain mathematical rigour. The fact is that, in the author's opinion, proofs of the convergence of algorithms do not consist solely of formal reasoning. They involve something more, namely an analysis of the reasons for the convergence and of those facts which may be an obstacle to this. Knowledge of these facts may form a basis for analysis of those cases where the algorithm does not lead to a successful solution of a problem.

It is well known that even the best algorithm may be ruined by a poor computer implementation. Thus, it is extremely important to use a whole arsenal of techniques for solving quadratic programming problems when solving the auxiliary problem (1.4), since solution of this problem requires the most machine time. In Section 1.3, we describe an algorithm for solving quadratic programming problems, which generalizes the simplex (linear programming) method in a multiplicative form. Use of this algorithm enables us to apply many modern techniques involving storage and computational economy and sparse-matrix operations. However, because of the specific form of (1.4), it is more convenient to go over to the dual problem which in this case, includes simple constraints on the variables and may be solved by a generalized conjugate gradient method widely used to minimize quadratic functions without constraints.

1.1.4 Notation

The book uses standard notation; however, to avoid misunderstanding, we shall describe this briefly here.

Everything takes place in the n-dimensional space $\mathbb{R}^n$, with column vectors x, y, z etc. Matrices are denoted by capital letters A, B, C and transposition is shown using an asterisk superscript. I_n denotes the unit matrix of order n. As usual, (x, y) denotes the scalar product of the vectors x and y:

$$(x,y)=\sum_{i=1}^{n} x^i y^i,$$

where the components of the vectors are denoted by superscripts. The Euclidean norm is used throughout, i.e.

$$\|x\|=\left(\sum_{i=1}^{n}(x^i)^2\right)^{1/2}.$$

We use $\overline{\lim}$ ($\underline{\lim}$) or lim sup (lim inf) to denote the upper (lower) limit. The notation $\lambda \downarrow 0$ means that λ is monotonically decreasing and tends to zero.

The derivatives of functions are denoted as follows. The gradient $f'(x)$ is a row vector with components $\partial f(x)/\partial x^i$, $i=1,\dots n$. $f''(x)$ denotes the matrix of second derivatives, i.e.

$$f''(x)=\left\{\frac{\partial^2 f(x)}{\partial x^i \partial x^j}\right\}_{\substack{i=1,\dots,n\\ j=1,\dots,n}}$$

Sometimes, when the argument with respect to which a derivative is taken is unclear, this argument may be written as a subscript. Thus, if

$$L(x,u)=f_0(x)+\sum_{i=1}^{n} u^i f_i(x)$$

is the Lagrange function, $L''_{xx}(x,u)$ denotes the matrix of second derivatives with respect to x.

Finally, $\min_x\{f(x)|x\in M\}$ denotes the minimum value of the function f as the argument x ranges over the set M.

1.2 Necessary Conditions for a Minimum and Duality

Current methods of solving mathematical programming problems are based on a theoretical foundation, involving convex analysis and the theory of necessary conditions for extrema. This section is intended to provide a short summary of the results which will be used at some stage in what follows.

Thus, this section is a self-contained handbook of the theory of convex analysis, duality and necessary conditions for extrema. While readers who are only interested in the practical side of things (i.e. the computational algorithms) may initially omit this section, turning to it as a handbook when necessary, they may nevertheless be interested in this condensed statement of the theoretical material on which current computational methods are based. More detailed studies of the material in this section may be found in [7,23,29,47].

1.2.1 Convex Sets

The set M in the space $\mathbb{R}^n$ is said to be *convex* if, given any two points $x_1, x_2 \in M$, M contains the whole of the segment joining them, i.e. $\lambda_1 x_1 + \lambda_2 x_2 \in M$, for all $\lambda_1, \lambda_2 \geq 0$, $\lambda_1 + \lambda_2 = 1$. This property, which characterizes a convex set, is easy to generalize. If M is convex, then given $x_i \in M$, $i = 1, \dots, m$, we have $\lambda_1 x_1 + \dots + \lambda_m x_m \in M$, for all $\lambda_i \geq 0$, $\lambda_1 + \dots + \lambda_m = 1$.

It follows directly from the definitions of the corresponding concepts that the interior $\operatorname{int} M$ and the closure $\operatorname{cl} M$ of a convex set M are also convex.

One very important property of convex sets is the fact that a point which does not belong to a convex set may be separated from that set. More precisely, if M is a closed convex set and the point x_0 does not belong to it, then there exist a vector $a \in \mathbb{R}^n$ and $\epsilon > 0$ such that

$$(x, a) \leq (x_0, a) - \epsilon$$

for all $x \in M$. In constructing dual algorithms, we often use the fact that the vector a may be chosen to be the vector $a = x_0 - y$ where y is the point of the set M which is closest to x_0, i.e.

$$\|x_0 - y\| = \min_x \{\|x_0 - x\| | x \in M\}, \quad y \in M.$$

Convex cones are convex sets with particular features. A convex set K is said to be a *convex cone* if whenever $x \in K$, it follows that $\lambda x \in K$ for all $\lambda > 0$.

Each convex cone is closely associated with its dual cone K^*. By definition:

$$K^* = \{y \in \mathbb{R}^n | (x, y) \geq 0 \ \ \forall x \in K\}. \tag{1.7}$$

If we associate each $y \in \mathbb{R}^n$ with the linear function (x, y), then the *dual cone* K^* may be viewed as the set of linear functions taking nonnegative values on the cone K.

A typical example of a cone is a set defined by a system of linear inequalities:

$$K = \{x | (a_i, x) \geq 0, \quad i = 1, \dots, m\}.$$

It is known that the dual cone K^* consists of elements y which may be represented in the form

$$y = \sum_{i=1}^{m} u^i a_i, \quad u^i \geq 0, \quad i = 1, \dots, m.$$

1.2.2 Convex Functions

We shall consider functions defined for $x \in \mathbb{R}^n$ with values on the extended real axis. Thus, $-\infty$ and $+\infty$ are permissible values for $f(x)$. This convention about the values of the function f is convenient when considering dual problems in convex programming.

We associate each function f with two sets: namely its *epigraph* epi f, which is the set of pairs $\{x, \alpha\}$ such that $x \in \mathbb{R}^n$, $\alpha \in \mathbb{R}$ and $\alpha \geq f(x)$ and its *domain* dom f, which is the set of x such that $f(x) < +\infty$.

Thus, by definition

$$\begin{aligned}\operatorname{epi} f &= \{\{x, \alpha\} \in \mathbb{R}^{n+1} | \alpha \geq f(x)\} \\ \operatorname{dom} f &= \{x \in \mathbb{R}^n | f(x) < +\infty\}.\end{aligned}$$

A general *convex function* is defined to be a function for which the corresponding epigraph is a convex set. It is easy to see that the domain of a convex function is a convex set.

In what follows, we shall consider only nonsingular convex functions, i.e. those which do not take the value $-\infty$ and are not identically equal to $+\infty$. Here we note the following fact. Suppose that $f(y) = -\infty$ at some point y and that $x \in \operatorname{int} \operatorname{dom} f$, where f is a convex function. Then for a sufficiently small $\epsilon > 0$, we have $x_1 = x + \epsilon(x - y) \in \operatorname{dom} f$. It is easy to see that

$$x = \frac{1}{1+\epsilon} x_1 + \frac{\epsilon}{1+\epsilon} y.$$

Since $f(y) = -\infty$, it follows that $\{y, \beta\} \in \operatorname{epi} f$ for arbitrary β. Suppose that $\alpha_1 \geq f(x_1)$, i.e. $\{x_1, \alpha_1\} \in \operatorname{epi} f$. Since the epigraph is convex, we have

$$\left\{\frac{1}{1+\epsilon} x_1 + \frac{\epsilon}{1+\epsilon} y, \frac{1}{1+\epsilon} \alpha_1 + \frac{\epsilon}{1+\epsilon} \beta\right\} \in \operatorname{epi} f,$$

i.e.

$$f(x) = f\left(\frac{1}{1+\epsilon} x_1 + \frac{\epsilon}{1+\epsilon} y\right) \leq \frac{1}{1+\epsilon} \alpha_1 + \frac{\epsilon}{1+\epsilon} \beta.$$

Since β is arbitrary, it follows that $f(x) = -\infty$. Thus, if the convex function f takes the value $-\infty$, then $f(x) = -\infty$ for all $x \in \operatorname{int} \operatorname{dom} f$.

For nonsingular convex functions, to which we shall restrict our attention in what follows, the definition in terms of the convexity of the epigraph is equivalent to the usual definition: the function f is said to be *convex* if

$$f(\lambda_1 x_1 + \lambda_2 x_2) \leq \lambda_1 f(x_1) + \lambda_2 f(x_2), \quad \lambda_1, \lambda_2 \geq 0, \quad \lambda_1 + \lambda_2 = 1. \tag{1.8}$$

The convexity of a function implies a number of analytical properties. For example, a necessary and sufficient condition for a convex function to be continuous at the point x_0 is that it should be bounded above in some neighbourhood of that point. Moreover, in this case, it satisfies a Lipschitz condition. If a function has first or second derivatives, the inequalities

$$\begin{aligned}f(y) - f(x) &\geq (y - x, f'(x)) \ \forall y, x \\ (f''(x)p, p) &\geq 0 \ \forall p, x\end{aligned}$$

are equivalent to convexity. In particular, if $f(x) = 1/2(x, Cx) + (d, x)$, then f is convex if and only if $(p, Cp) \geq 0$, i.e. the matrix C is positive semi-definite.

Generally speaking, a convex function is not smooth and does not have continuous derivatives. However, its *directional derivative*

$$f'(x,p) = \lim_{\lambda \downarrow 0} \frac{f(x+\lambda p) - f(x)}{\lambda}$$

exists and is finite at any point at which f is continuous. Here, the difference quotient $\lambda^{-1}[f(x+\lambda p) - f(x)]$ is monotonic decreasing and tends to $f'(x,p)$.

We introduce the following definition for a convex function. The vector x^* is said to be a *subgradient* of the function f at the point x if

$$f(y) - f(x) \geq (y - x, x^*)$$

for all y. The set of all subgradients is denoted by $\partial f(x)$ and is called the *subdifferential.* Thus,

$$\partial f(x) = \{x^* \in \mathrm{I\!R}^n | f(y) - f(x) \geq (y - x, x^*) \ \ \forall y\}. \tag{1.9}$$

If the function is differentiable, then the subdifferential consists of a single vector $f'(x)$.

Subdifferentials and directional derivatives are closely related. Thus, if f is continuous at the point x, then

$$f'(x,p) = \max_{x^*}\{(p, x^*) | x^* \in \partial f(x)\} \tag{1.10}$$

and $\partial f(x) \neq \emptyset$.

We shall say that the function f is *closed* if its epigraph is a closed set. The following three statements are equivalent:

1. The function f is closed.
2. The sets $\{x | f(x) \leq \alpha\}$ are closed for all α.
3. The function f is lower semicontinuous, i.e,
$$\liminf_{y \to x} f(y) \geq f(x).$$

Just as there is a close connection between the convex cone and the dual cone, in the same way, for a convex function, we may define the *dual function*

$$f^*(x^*) = \sup_x\{(x, x^*) - f(x)\}. \tag{1.11}$$

This function is always convex and closed. The connection between the original and the dual function is established by the following important theorem.

Theorem 1.1 *Suppose that f is a nonsingular convex function which is lower semicontinuous at the point x. Then*

$$f(x) = f^{**}(x),$$

where

$$f^{**}(x) = \sup_{x^*}\{(x, x^*) - f^*(x^*)\}.$$

This theorem has numerous applications. In particular, as will be clear from the next subsection, it forms the basis for the theory of duality in convex programming.

1.2.3 Foundations of Convex Programming

Convex programming problems involve the minimization of a convex function on a convex set. Necessary conditions for extrema may be given, depending on the form of the representation of the convex set.

To begin with, we consider a very simple problem. Suppose that f is a smooth convex function and that M is a convex set. We wish to determine a point x_0 at which f attains its minimum on M.

Let

$$K_M(x_0) = \{p | p = \lambda(x - x_0), \ \ \lambda > 0, \ x \in M\}.$$

If $p \in K_M(x_0)$, then

$$x_0 + \alpha p = (1 - \alpha\lambda)x_0 + \alpha\lambda x \in M$$

for $\alpha \leq \lambda^{-1}$. Thus, for small displacements from x_0 in the direction p, the point $x_0 + \alpha p$ does not lie outside the set M. Thus, the cone $K_M(x_0)$ is called the *cone of feasible directions.*

Theorem 1.2 *A necessary and sufficient condition for a point x_0 to be a minimum of the smooth convex function f on the convex set M is that $f'(x_0) \in K_M^*(x_0)$.*

Proof. Suppose that x_0 is a minimum of f. Then for any $x \in M$ and $0 < \lambda \leq 1$,

$$f((1 - \lambda)x_0 + \lambda x) = f(x_0 + \lambda(x - x_0)) \geq f(x_0),$$

or

$$\frac{f(x_0) + \lambda p) - f(x_0)}{\lambda} \geq 0.$$

Passing to the limit, we obtain:

$$f'(x_0, p) = (p, f'(x_0)) \geq 0, \ \ p = x - x_0,$$

whence

$$(p, f'(x_0)) \geq 0, \ \ p \in K_M(x_0). \tag{1.12}$$

From the definition of the dual cone, the last inequality implies that $f'(x_0) \in K_M^*(x_0)$.

Conversely, if (1.12) is satisfied, then because f is convex

$$f(x) - f(x_0) \geq (x - x_0, f'(x_0)) \geq 0, \ \ x \in M,$$

i.e. x_0 is a minimum of f. □

Let us consider the case where the set M is given by a system of linear inequalities:

$$M = \{x | (a_i, x) \leq \alpha_i, \ \ i = 1, \ldots, m\}. \tag{1.13}$$

We set

$$I(x) = \{i|(a_i, x) = \alpha_i, \ \ i = 1, \dots, m\}. \tag{1.14}$$

It is easy to see that only directions p which satisfy the inequalities $(a_i, p) \le 0$, $i \in I(x_0)$, have the property that for a small displacement from the point x_0 in the direction p, the point $x_0 + \alpha p$ remains in M. Thus,

$$K_M(x_0) = \{p|(a_i, p) \le 0, \ \ i \in I(x_0)\}.$$

Using the statement in Section 1.2.1 about the dual cone of a cone given by a system of linear inequalities, we obtain that $y \in K_M^*(x_0)$ if and only if

$$y = -\sum_{i \in I(x_0)} u^i a_i, \ \ u^i \ge 0, \ \ i \in I(x_0).$$

Thus, if x_0 is a minimum of a smooth convex function f on the set M given by (1.13), then

$$f'(x_0) = -\sum_{i \in I(x_0)} u^i a_i, \ \ u^i \ge 0, \ \ i \in I(x_0). \tag{1.15}$$

We give this result in a somewhat different form.

Theorem 1.3 *The point x_0 is a minimum of the smooth convex function f with the constraints*

$$(a_i, x) \le \alpha_i, \ \ i = 1, \dots, m$$

if and only if there exist numbers $u^i \ge 0$, $i = 1, \dots, m$, such that

$$\begin{aligned} f'(x_0) + \sum_{i=1}^{m} u^i a_i &= 0 \\ u^i[(a_i, x) - \alpha_i] &= 0, \ \ i = 1, \dots, m. \end{aligned} \tag{1.16}$$

Proof. The proof follows from the above, if we set $u^i = 0$, $i \notin I(x_0)$ and recall the definition (1.14) of the set $I(x_0)$. □

Remark. This result is easy to generalize to the case where the definition of the set M includes equations $(a_i, x) = \alpha_i$. Of course, such an equation is equivalent to two inequalities $(a_i, x) \le \alpha_i$ and $(-a_i, x) \le -\alpha_i$. Thus, (1.16) will include vectors a_i and $-a_i$ with factors $u_+^i \ge 0$ and $u_-^i \ge 0$, respectively. Collecting similar terms and denoting $u^i = u_+^i - u_-^i$, we obtain the same equation (1.16), except that the sign of u^i is no longer constrained.

Using this remark, it is not difficult to obtain necessary and sufficient conditions for the following standard form of the problem.

Theorem 1.4 *The point x_0 is a minimum of the smooth convex function f with the constraints*

$$\begin{aligned}(a_i, x) &= \alpha_i, \;\; i = 1, \dots m, \\ x^j &\geq 0, \;\; j = 1, \dots, n\end{aligned}$$

if and only if there exist numbers u^i and a vector $v \in \mathbb{R}^n$ with components v^j, such that

$$\begin{aligned}f'(x_0) + \sum_{i=1}^{m} u^i a_i &= v \\ v^j \geq 0, \;\; v^j x_0^j &= 0, \;\; j = 1, \dots, n.\end{aligned}$$

Proof. The proof is obtained directly from Theorem 1.3 and the remark following it, taking into account the fact that the inequality $x^j \geq 0$ is equivalent to the inequality $(-e_j, x) \leq 0$, where $e_j^* = (0, 0, \dots, 0, 1, 0, \dots, 0)$ is the jth unit vector. Here, v^j is the factor corresponding to this constraint. □

Suppose now that $f_i(x)$, $i = 0, 1, \dots, m$ are convex continuous functions and that M is a convex set. Let us consider the problem $P(0)$:

$$\min_x \{f_0(x) | f_i(x) \leq 0, \; i = 1, \dots, m, \; x \in M\}.$$

For further convenience, this problem is embedded in a class of minimization problems $P(y)$, depending on a parameter vector $y \in \mathbb{R}^m$. We set

$$V(y) = \inf_x \{f_0(x) | f_i(x) \leq y^i, \; i = 1, \dots, m, \; x \in M\}.$$

By convention, we set $V(y) = +\infty$, if for a given y the constraints of the problem are inconsistent. Problem $P(y)$ is defined to be that of minimizing the right-hand side of the above equation. The original problem $P(0)$ is obtained for $y = 0$.

Lemma 1.1 *The function $V(y)$ is convex. If there exists a point $x \in M$ such that*

$$f_i(\overline{x}) < 0, \;\; i = 1, \dots, m,$$

and $V(0)$ is finite, then $V(y)$ is continuous in a neighbourhood of the point $y = 0$ and $\partial V(0) \neq \emptyset$.

Proof. Suppose that $\beta_j > V(y_j)$, $j = 1, 2$. Then there exist points $x_j \in M$ such that

$$f_0(x_j) < \beta_j, \;\; f_i(x_j) \leq y_j^i, \;\; j = 1, 2.$$

Since the functions involved are convex, the following inequalities are satisfied:

$$\begin{aligned}f_0(\lambda_1 x_1 + \lambda_2 x_2) &= \lambda_1 f_0(x_1) + \lambda_2 f_0(x_2) < \lambda_1 \beta_1 + \lambda_2 \beta_2 \\ f_i(\lambda_1 x_1 + \lambda_2 x_2) &\leq \lambda_1 y_1^i + \lambda_2 y_2^i, \;\; i = 1, \dots, m \\ \lambda_1 x_1 + \lambda_2 x_2 &\in M, \;\; \lambda_1 + \lambda_2 = 1, \;\; \lambda_1, \lambda_2 \geq 0.\end{aligned}$$

Thus,

$$V(\lambda_1 y_1 + \lambda_2 y_2) < \lambda_1 \beta_1 + \lambda_2 \beta_2,$$

whence it follows that $\operatorname{epi} V$ is a convex set and thus that the function V is convex.

If there exists a point $\overline{x}$ as described in the lemma, then clearly $f_i(\overline{x}) \leq y^i$, $i = 1, \dots, m$ for small y and thus $V(y) \leq f_0(\overline{x})$ and $0 \in \operatorname{int} \operatorname{dom} V$. Since $V(0)$ is finite, following Section 1.2.2, V cannot have value $-\infty$.

Thus, the function V is convex and finite in a neighbourhood of zero and bounded above in this neighbourhood by the number $f_0(\overline{x})$. Following Section 1.2.2, $V(y)$ is continuous in a neighbourhood of zero and $\partial V(0) \neq \emptyset$. This completes the proof of the lemma. □

The vector $u \in \mathbb{R}^m$ is called a *Kuhn–Tucker vector* of the convex programming problem if $u \geq 0$ and

$$V(0) = \inf_x \{L(x, u) | x \in M\},$$

where

$$L(x, u) = f_0(x) + \sum_{i=1}^{\infty} u^i f_i(x).$$

The function L is called the *Lagrange function.*

Theorem 1.5 *Suppose that $V(0)$ is finite. The vector u is a Kuhn–Tucker vector if and only if $-u \in \partial V(0)$.*

Proof. By the definition of the subdifferential, $-u \in \partial V(0)$ if and only if $V(y) \geq V(0) - (u, y)\ \forall y$, i.e.

$$\inf_y \{V(y) + (u, y)\} = V(0).$$

Substituting the expression for $V(y)$ here, we obtain

$$\inf_y \inf_x \{f_0(x) + (u, y) | f_i(x) \leq y^i,\ i = 1, \dots, m,\ x \in M\} = V(0).$$

It is easy to see that, if for some i the corresponding component u^i of the vector u is negative $u^i < 0$ then, letting y^i tend to $+\infty$, we obtain $-\infty$ on the left-hand side of this inequality, which contradicts the assumption that $V(0)$ is finite. Thus, $u \geq 0$ and the minimum of the left-hand side with respect to y is attained when $y^i = f_i(x)$.

Thus,

$$\inf_x \{f_0(x) + \sum_{i=1}^{m} u^i f_i(x) | x \in M\} = V(0),$$

i.e.

$$\inf_x \{L(x, u) | x \in M\} = V(0)$$

and the vector u is a Kuhn–Tucker vector. The theorem has been proved. $\square$

Theorem 1.6 *Suppose there exists a point $\overline{x} \in M$ such that $f_i(\overline{x}) < 0$ for $i = 1, \ldots, m$ and that $V(0)$ is finite. Then there exists a Kuhn–Tucker vector. If x_0 is a solution of problem $P(0)$, then*

$$\begin{aligned} f_0(x_0) &= L(x_0, u) \leq L(x, u), \quad x \in M, \\ u^i &\geq 0, \quad u^i f_i(x_0) = 0, \quad i = 1, \ldots, m. \end{aligned} \tag{1.17}$$

Conversely, if the point x_0 satisfies the constraints of problem $P(0)$ and together with the vector u satisfies the condition (1.17), then x_0 is a solution of problem $P(0)$ and u is a Kuhn–Tucker vector.

Proof. The existence of a Kuhn–Tucker vector follows from Lemma 1.1 and Theorem 1.5. Suppose now that $x_0 \in M$ is a solution of problem $P(0)$. Then

$$\begin{aligned} f_i(x_0) &\leq 0, \quad i = 1, \ldots, m \\ f_0(x_0) &= V(0). \end{aligned} \tag{1.18}$$

Since $u \geq 0$, we have

$$V(0) = f_0(x_0) \geq f_0(x_0) + \sum_{i=1}^{m} u^i f_i(x_0) = L(x_0, u).$$

But, by the definition of a Kuhn–Tucker vector, $V(0) \leq L(x_0, u)$. Thus,

$$f_0(x_0) = L(x_0, u) = V(0) \leq L(x, u), \quad x \in M.$$

Moreover, from $u \geq 0$, (1.18) and the equation

$$f_0(x_0) = L(x_0, u) = f_0(x_0) + \sum_{i=1}^{m} u^i f_i(x_0),$$

it follows that

$$u^i f_i(x_0) = 0, \quad i = 1, \ldots, m.$$

We now prove the remaining part of the theorem. Suppose that the vectors u and x_0 satisfy the constraints of problem $P(0)$ and are such that equation (1.17) is satisfied. Let x be any point which satisfies the constraints of problem $P(0)$. Then, it follows from (1.17) that

$$f_0(x_0) = L(x_0, u) \leq L(x, u) = f_0(x) + \sum_{i=1}^{m} u^i f_i(x) \leq f_0(x),$$

i.e. x_0 is a solution of the problem. Thus, $V(0) = f_0(x_0)$ and, by virtue of (1.17), u is a Kuhn–Tucker vector. $\square$

Theorem 1.7 *Suppose that, for the conditions of Theorem 1.6, the functions f_i are continuously differentiable. Then, the point x_0 is a solution of problem*

$P(0)$ if and only if there exists a vector $u \in \mathbb{R}^m$ such that the following equations are satisfied:

$$\begin{aligned} L'_x(x_0, u) &\in K^*_M(x_0) \\ u \geq 0, \ u^i f_i(x_0) &= 0, \ i = 1, \dots, m \end{aligned} \tag{1.19}$$

Proof. In the present smooth situation, (1.19) is completely equivalent to (1.18). In fact, the first condition of (1.18) asserts that the point x_0 is a minimum of $L(x, u)$ over all $x \in M$. But, from Theorem 1.2, the first equation of (1.19) is a necessary and sufficient condition for this. All the remaining equations in (1.18) and (1.19) are the same. □

1.2.4 Duality in Convex Programming

The problem $P(0)$ considered in Section 1.2.3 may be associated with the dual problem. For this, we calculate the dual function of the function $V(0)$ of Section 1.2.3.

By definition,

$$\begin{aligned} V^*(u) &= \sup_y \{(u, y) - V(y)\} \\ &= \sup_y \{(u, y) - \inf_x \{f_0(x) | f_i(x) \leq y^i, \ i = 1, \dots, m, \ x \in M\}\} \\ &= \sup_x \sup_y \{-f_0(x) + \sum_{i=1}^{\infty} u^i y^i | f_i(x) \leq y^i, \ i = 1, \dots, m, \ x \in M\}. \end{aligned}$$

It is easy to compute the upper bound with respect to y. This gives:

$$V^*(u) = \begin{cases} \sup_x \{-f_0(x) + \sum_{i=1}^m u_i f_i(x) | x \in M\}, & u \leq 0, \\ +\infty & u^i > 0 \text{ for some } i. \end{cases}$$

In other words,

$$V^*(u) = \begin{cases} \inf_x \{L(x, -u) | x \in M\} & u \leq 0, \\ +\infty & u^i > 0 \text{ for some } i. \end{cases} \tag{1.20}$$

Theorem 1.8 *If the function $V(y)$ is lower semi-continuous for $y = 0$, and, in particular, if $V(0)$ is finite and there exists $x \in M$ such that $f_i(x) < 0$, $i = 1, \dots, m$, then*

$$V(0) = \sup_{u \geq 0} \inf_x \{L(x, u) | x \in M\}. \tag{1.21}$$

If there exists a Kuhn–Tucker vector then (1.21) is satisfied and the upper bound is attained by this vector.

Proof. It follows from Theorem 1.1 and (1.20) that

$$V(0) = V^{**}(0) = \sup_u \{(u, 0) - V^*(u)\} = \sup_{u<0} \{\inf_x L(x, -u) | x \in M\}.$$

Replacing u by $-u$, we obtain the first assertion of the theorem.

Suppose now that $u \geq 0$ is an arbitrary vector. We have

$$\begin{aligned}\inf_x\{L(x,u)|x \in M\} &= \inf_x\{f_0(x) + \sum_{i=1}^m u^i f_i(x)|x \in M\} \\ &\leq \inf_x\{f_0(x)|f_i(x) \leq 0,\ i = 1,\dots,m,\ x \in M\} = V(0).\end{aligned}$$

Thus, for an arbitrary $u \geq 0$,

$$\inf_x\{L(x,u)|x \in M\} \leq V(0).$$

But, if $u_0 \geq 0$ is a Kuhn–Tucker vector, then

$$V(0) = \inf_x\{L(x,u_0)|x \in M\}.$$

This completes the proof. □

Let us set

$$\varphi(u) = \inf_x\{L(x,u)|x \in M\}.$$

The problem $\sup_{u \geq 0} \varphi(u)$ is called the *dual problem of convex programming.*

Thus, Theorem 1.8 asserts that, under certain assumptions, the upper bound in the dual problem is equal to the value $V(0)$ of the original problem and that if a Kuhn–Tucker vector exists, then the upper bound in the dual problem is attained by that vector.

Let us now return to the problem of minimization of a smooth convex function f with linear constraints $(a_i, x) \leq \alpha_i$, $i = 1,\dots,m$. If the solution is x_0, then, according to Theorem 1.3, there exists a vector $u \geq 0$ such that equation (1.16) is satisfied. But, for the present problem

$$L(x,u) = f(x) + \sum_{i=1}^m u^i[(a_i,x) - \alpha_i], \tag{1.22}$$

and thus equation (1.16) agrees exactly with (1.19), taking into account that in this problem $M = \mathbb{R}^n$, whence $K_M(x_0) = \mathbb{R}^n$ and $K_M^*(x_0) = \{0\}$. Next, by virtue of the equivalence of (1.19) and (1.17), we deduce the following result from Theorem 1.6:

Theorem 1.9 *If in the problem of minimization of a smooth convex function with linear constraints a minimum is attained, then there exists a Kuhn–Tucker vector, formula (1.21) holds and the Kuhn–Tucker vector gives a solution of the dual problem.*

1.2.5 Necessary Conditions for Extrema. General Problem

We now abandon convexity conditions and consider a general mathematical programming problem. In what follows, we shall consider only smooth problems.

Thus, suppose that $f_0(x)$, $f_i(x)$, $i \in I$ are continuously differentiable functions and that M is a convex set. Let us consider the problem

$$\min_x \{f_0(x) | f_i(x) \leq 0, i \in I^-, \ f_i(x) = 0, \ i \in I^0, \ x \in M\}, \tag{1.23}$$

where I is a finite set of indices and $I^- \cup I^0 = I$ and $I^- \cap I^0 = \emptyset$.

Suppose that x_0 is a solution of problem (1.23). Then, the following result holds [23]:

Theorem 1.10 *A necessary condition for x_0 to be a solution of problem (1.23) is that there should exist numbers u^i, $i \in \{0\} \cup I$, not all simultaneously zero, such that*

$$\text{(i)} \qquad u^0 f_0'(x_0) + \sum_{i \in I} u^i f_i'(x_0) \in K_M^*(x_0)$$

$$\text{(ii)} \qquad u^i \geq 0, \ \ i \in \{0\} \cup I^-, \ \ u^i f_i(x_0) = 0, \ \ i \in I^-. \tag{1.24}$$

The values u^i in the equations (1.24) are called *Lagrange multipliers.*

Remark. If $M = \mathbb{R}^n$, then $K_M^*(x_0) = \{0\}$ and thus condition (i) of (1.24) becomes the equation

$$u^0 f_0'(x_0) + \sum_{i \in I} u^i f_i'(x_0) = 0. \tag{1.25}$$

In condition (1.24)(i), the gradient of the function $f_0(x)$ is multiplied by u^0. If this coefficient is zero, the problem becomes degenerate, since the function to be minimized plays practically no role in the necessary conditions for extrema. Thus, the following statement identifying the regular case is reasonable. The minimum x_0 in problem (1.23) is *regular* if the number u^0 in (1.24)(i) is strictly positive.

Remark. It is easy to show that, in this case, by dividing the whole of equation (1.24) by u^0, we may assume that $u^0 = 1$. We shall make this assumption.

If the minimum x_0 is regular, then condition (1.24)(i) may be rewritten in the form

$$L_x'(x_0, u) \in K_M^*(x_0), \tag{1.26}$$

where $L(x, u)$ is the Lagrange function. Comparing (1.26), (1.24) and (1.17), we see that Theorems 1.6 and 1.7 considered the regular case of the convex programming problem.

In the general case, whilst regularity conditions are extremely important as far as constructing algorithms is concerned, they are quite difficult to formulate in a readily verifiable form. As can be seen from the above, if the functions f_i are convex and $I^0 = \emptyset$, then such a condition is the existence of a point $\overline{x} \in M$ such that $f_i(\overline{x}) < 0$ for $i \in I^-$. If the f_i are not assumed to be convex, then it is only possible to formulate a far weaker condition.

Suppose that $M = \mathbb{R}^n$. The minimum x_0 of problem (1.23) is said to be *strongly regular*, if the vectors $f_i'(x_0)$ are linearly independent, $i \in I^-(x_0) \cup I^0$, where $I^-(x_0) = \{i \in I | f_i(x_0) = 0\}$.

If we take into account (1.24)(ii), from which it follows that $u^i = 0$ for $i \notin I^-(x_0)$, then, from (1.25), the strong regularity condition implies that $u^0 > 0$, since the equation with $u^0 = 0$ would at the same time be a linear relationship between the vectors $f_i'(x_0)$, $i \in I^-(x_0) \cup I^0$, which would contradict the strong regularity.

Equation (1.25) may be rewritten in the form

$$f_0'(x_0) + \sum_{i \in I^-(x_0) \cup I^0} u^i f_i'(x_0) = 0.$$

Since the vectors $f_i'(x_0)$, $i \in I^-(x_0) \cup I^0$ are linearly independent, the last equation determines the u^i uniquely. Thus, the following theorem holds.

Theorem 1.11 *Suppose that the minimum x_0 of problem (1.23) is strongly regular. Then it is regular, i.e. u^0 in (1.24) may be set equal to 1 and all the other values u^i are uniquely determined.*

1.2.6 Necessary Conditions for Extrema: Second Order Conditions

The proofs of the results given in the previous section use various arguments assuming first order smallness; thus, the statements of the theorems only involve the first derivatives of the original functions. Study of higher order variants leads to necessary conditions involving second order derivatives. In what follows, we shall require the following result.

Suppose that $M = \mathbb{R}^n$ and that the functions f_i are twice continuously differentiable.

Theorem 1.12 *Suppose that the minimum x_0 in problem (1.23) is strongly regular and that $u^i > 0$, $i \in I^-(x_0)$. Then*

$$(L''_{xx}(x_0, u)p, p) \geq 0 \tag{1.27}$$

for all $p \in \mathbb{R}^n$ satisfying the condition

$$f_i'(x_0)p = 0, \quad i \in I^-(x_0) \cup I^0.$$

1.2.7 Minimax Problems

Suppose that we require to find a minimum of the function

$$f_0(x) = \max_{k \in I} \varphi_k(x) \tag{1.28}$$

where I is a finite set of indices and the functions φ_k are continuously differentiable. Here, x varies throughout the space $\mathbb{R}^n$. It is easy to reduce this problem to a general mathematical programming problem by introducing an additional variable x^0. Thus, the problem of minimizing the function $f(x)$ given by formula (1.28), is equivalent to the problem

$$\min_{\{x,x^0\}} \{x^0 | \varphi_k(x) \leq x^0, k \in I\}. \tag{1.29}$$

Applying Theorem 1.10 to this problem (for $M = \mathbb{R}^{n+1}$), leads to the following result.

Theorem 1.13 *For the point x_0 to be a minimum of the function $f_0(x)$ of equation (1.28), there must exist numbers u^k, $k \in I$ such that*

$$\begin{aligned} \sum_{k\in I} u^k \varphi_k'(x_0) &= 0, \\ u^k \geq 0, \;\; u^k(\varphi_k(x_0) - f_0(x_0)) &= 0, \\ \sum_{k\in I} u^k &= 1 \end{aligned}$$

It is easy to see that if x is constrained by constraints of the form (1.23), determined by the conditions for which it is required to minimize the function $f_0(x)$, then, the problem may be reduced to a general mathematical programming problem in a similar way. However, in this case, there are no specific features which distinguish it from the general problem, and thus, we shall merely observe the above fact here.

1.2.8 Penalty Functions

The method of penalty functions is quite popular for solving mathematical programming problems. It reduces such problems to problems of unconstrained optimization. There is a vast amount of literature about the method of penalty functions to which readers desiring more detailed information should refer. Here, we shall only study a number of facts relating to the linearization method.

Suppose that $f_0, f_1, \ldots, f_m$ are continuous functions and that M is some set. We set

$$V(y) = \inf_x \{f_0(x) | f_i(x) \leq y^i, \; i = 1, \ldots, m, \; x \in M\}. \tag{1.30}$$

The above optimization problem (1.30) will be denoted by $P(y)$. Of course, in the first place, we shall be interested in the initial problem $P(0)$.

We also note that the fact that (1.30) only includes constraints in the form of inequalities does not decrease the generality, since each constraint in the form of an equation $f(x) = 0$ may be written as two inequalities: $f(x) \leq 0$ and $-f(x) \leq 0$.

We set

$$\begin{aligned} F(x) &= \max\{0, f_1(x), \ldots, f_m(x)\} \\ \Phi_N(x) &= f_0(x) + NF(x). \end{aligned}$$

The following chain of equations is immediately apparent:

$$\begin{aligned}\inf_{x\in M} \Phi_N(x) &= \inf_{x,\lambda}\{f_0(x) + N\lambda | 0 \le \lambda, f_1(x) \le \lambda, \dots, f_m(x) \le \lambda, x \in M\} \\ &= \inf_{\lambda\ge 0}[V(\lambda\cdot\mathbf{1}) + N\lambda], \quad \mathbf{1} = \begin{pmatrix} 1 \\ \vdots \\ 1 \end{pmatrix} \in \mathbb{R}^m. \end{aligned} \tag{1.31}$$

Whence it follows that if the lower bound on the right-hand side of (1.31) is only attained for $\lambda = 0$, i.e.

$$V(\lambda\cdot\mathbf{1}) - V(0) > N\lambda, \quad \lambda > 0, \tag{1.32}$$

then

$$V(0) = \inf_x\{\Phi_N(x) | x \in M\}. \tag{1.33}$$

Thus, if (1.32) is satisfied, the lower bound in problem $P(0)$ coincides with the lower bound of the function $\Phi_N(x)$ for $x \in M$. We shall show that a stronger results holds, namely that points at which the minimum of problem $P(0)$ is attained coincide with the minima of $\Phi_N(x)$.

Theorem 1.14 *Suppose that*

$$\inf_{\lambda>0} \frac{V(\lambda\cdot\mathbf{1}) - V(0)}{\lambda} = -L > -\infty \tag{1.34}$$

and $N > L$. Then the minima of problem $P(0)$ and those of the problem $\inf_x\{\Phi_N(x)|x \in M\}$ coincide.

Remark. Since $V(\lambda\cdot\mathbf{1})$ is clearly a decreasing function of λ (the lower bound in a narrower region is greater than the lower bound in a wider region), we have $L \ge 0$.

Proof. It follows from (1.34) that $V(\lambda\cdot\mathbf{1}) + N\lambda > V(0)$ for $\lambda > 0$.

Suppose that $x_0 \in M$ and $\Phi_N(x_0) = \inf_x\{\Phi_N(x)|x \in M\}$. Then

$$\inf_x\{f_0(x) + NF(x)|x \in M\} \le \inf_x\{f_0(x)|f_i(x) \le 0, \ i = 1, \dots, m, \ x \in M\}$$

i.e.

$$V(0) \ge \inf_x\{\Phi_N(x)|x \in M\}. \tag{1.35}$$

We shall show that $F(x_0) = 0$. Let us assume the contrary, i.e. that $\lambda_0 = F(x_0) > 0$. We have

$$\begin{aligned} f_0(x_0) + NF(x_0) &= \min_x\{f_0(x) + NF(x)|x \in M\} \\ &\le \min_x\{f_0(x) + NF(x)|x \in M, \ F(x) \le \lambda_0\} \\ &\le \min_x\{f_0(x) + N\lambda_0|x \in M, \ F(x) \le \lambda_0\} = V(\lambda_0\cdot\mathbf{1}) + N\lambda_0. \end{aligned}$$

But, on the other hand,

$$\begin{aligned} f_0(x_0) + NF(x_0) &= f_0(x_0) + N\lambda_0 \\ &\geq \min_x\{f_0(x) + N\lambda_0 | x \in M, F(x) \leq \lambda_0\} \\ &\geq V(\lambda_0 \cdot \mathbf{1}) + N\lambda_0. \end{aligned}$$

Thus,

$$f_0(x_0) + NF(x_0) = V(\lambda_0 \cdot \mathbf{1}) + N\lambda_0.$$

Whence, from (1.34) and (1.35) and the fact that $N > L$, we obtain

$$V(0) \geq f_0(x_0) + NF(x_0) = V(\lambda_0 \cdot \mathbf{1}) + N\lambda_0 > V(0),$$

since $\lambda_0 = F(x_0) > 0$. We have obtained a contradiction.

Thus, $F(x_0) = 0$, whence, it follows that the point $x_0 \in M$ satisfies all the constraints of problem $P(0)$. Thus, $f_0(x_0) \geq V(0)$. But, according to (1.35), $V(0) \geq f_0(x_0) + NF(x_0) = f_0(x_0)$. Finally, $f_0(x_0) = V(0)$, i.e. x_0 is a minimum for problem $P(0)$.

Conversely, suppose that x_0 is a solution of problem $P(0)$. Then

$$\begin{aligned} f_0(x_0) + NF(x_0) &= f_0(x_0) = V(0) \\ &\leq \inf_{\lambda \geq 0}[V(\lambda \cdot \mathbf{1}) + N\lambda] \\ &= \inf_x\{\Phi_N(x) | x \in M\} \end{aligned}$$

i.e. x_0 minimizes $\Phi_N(x)$ over M. □

A number of corollaries follow from the above result.

Corollary 1 *Suppose that the functions $f_i(x)$ $i = 0, 1, \ldots, m$ and the set M are convex and that there exists a Kuhn–Tucker vector u. Then, for*

$$N > \sum_{i=1}^{m} u^i$$

the solutions of problem $P(0)$ and the minima of $\Phi_N(x)$ over M coincide.

Proof. Since this problem is a convex programming problem, $V(y)$ is a convex function and, by virtue of Theorem 1.5, $-u \in \partial V(0)$. Thus,

$$\begin{aligned} V(\lambda \cdot \mathbf{1}) - V(0) &\geq -(u, \lambda \cdot \mathbf{1}) = -\lambda \sum_{i=1}^{m} u^i, \\ \inf_{\lambda \geq 0} \frac{V(\lambda \cdot \mathbf{1}) - V(0)}{\lambda} &\geq -\sum_{i=1}^{m} u^i, \end{aligned}$$

and the assertion follows from the theorem. □

Corollary 2. *Suppose that in the general problem, $M \in \mathbb{R}^n$, all the functions are differentiable and equation (1.34) is satisfied. Then any minimum is a regular point.*

Proof. From Theorem 1.14, it follows that if x_0 is a minimum of problem $P(0)$, then it is a minimum of the function $\Phi_N(x)$ for N sufficiently large. But

$$\Phi_N(x) = \max\{\varphi_i(x) | i = 0, 1, \dots, m\}$$

where

$$\varphi_0(x) = f_0(x), \quad \varphi_i(x) = f_0(x) + N f_i(x), \quad i = 1, \dots, m.$$

Thus, the problem of minimizing $\Phi_N(x)$ is a minimax problem to which Theorem 1.13 may be applied. Therefore, there exist numbers $u^i \geq 0$, $i = 0, 1, \dots, m$, such that

$$\begin{aligned} u^0 f_0'(x_0) + \sum_{i=1}^{m} u^i (f_0'(x_0) + N f_i'(x_0)) &= 0 \\ u^0 (f_0(x_0) - \Phi_N(x_0)) &= 0 \\ u^i (f_0(x_0) + N f_i(x_0) - \Phi_N(x_0)) &= 0, \quad i = 1, \dots, m \\ u^0 + \sum_{i=1}^{m} u^i &= 1. \end{aligned}$$

But

$$\Phi_N(x_0) = f_0(x_0) + N F(x_0) = f_0(x_0),$$

since x_0 is a solution of problem $P(0)$. Thus, if we denote $\overline{u}^i = N u^i$, $i = 1, \dots, m$, then the previous equations may be rewritten in the form

$$\begin{aligned} f_0'(x_0) + \sum_{i=1}^{m} \overline{u}^i f_i'(x_0) &= 0 \\ \overline{u}^i f_i(x_0) = 0, \quad \overline{u}^i &= 0, \quad i = 1, \dots, m. \end{aligned}$$

These are the standard necessary conditions for a minimum, in which $\overline{u}^0 = 1$, which corresponds to the regular case.

1.3 Quadratic Programming Problems

Quadratic programming problems involve finding the minimum of a quadratic function with a positive-definite matrix with linear constraints on the arguments. They are of interest in themselves, in particular since they represent a generalization of linear programming problems; in the context of this book, they are important since they occur as fundamental auxiliary problems when solving general nonlinear problems. Calculations have shown that the overall effectiveness of a nonlinear programming algorithm depends on the effectiveness of the algorithm used to solve the quadratic programming problem.

There are many different quadratic programming algorithms. In what follows, the most important such algorithms are finite algorithms, in other words, those which give a solution after a finite number of steps. This is because a large number of quadratic programming problems have to be solved as part of the general linearization algorithm, and thus it is not possible to apply procedures of infinite duration within the overall process.

1.3.1 Conjugate Gradient Method

We begin with the simplest problem, in other words, that of minimizing the function

$$f(x) = 1/2(x, Cx) + (d, x), \tag{1.36}$$

where C is a symmetric positive definite matrix, i.e. $(p, Cp) \geq 0$ for all $p \in \mathbb{R}^n$.

The vectors p_i, $i = 1, \ldots, n$, are said to be *conjugate with respect to the matrix* C if they are linearly independent and

$$(p_i, Cp_j) = 0, \quad i \neq j. \tag{1.37}$$

Conjugate vectors always exist, although they are not uniquely determined. It is known from linear algebra that a symmetric matrix has n orthogonal eigenvectors s_i, corresponding to eigenvalues λ_i, i.e.

$$\begin{aligned} Cs_i &= \lambda_i s_i, \quad i = 1, \ldots, n \\ (s_i, s_j) &= \begin{cases} 1 & i = j \\ 0 & i \neq j \end{cases}. \end{aligned}$$

Thus,

$$(s_i, Cs_j) = \lambda_j (s_i, s_j) = \begin{cases} \lambda_j & i = j \\ 0 & i \neq j \end{cases}$$

and the vectors s_i are conjugate. For a positive-definite matrix $\lambda_i \geq 0$ although we may have $\lambda_i = 0$ for some i; thus, generally speaking, the number (s_i, Cs_i) may be equal to zero.

When we know a conjugate system of vectors, the minimization problem (1.36) is easy to solve. Suppose that x_0 is an arbitrary point. Any other arbitrary point x may be uniquely represented in the form

$$x = x_0 + \sum_{i=1}^{n} \alpha_i p_i$$

Substituting this expression in (1.36), we obtain:

$$f(x) = f(x_0) + \frac{1}{2} \sum_{i=1}^{n} \alpha_i^2 (p_i, Cp_i) + \sum_{i=1}^{n} \alpha_i (p_i, f'(x_0)), \tag{1.38}$$

where it is understood that $f'(x) = Cx + d$. Then it is easy to find a minimum of the right-hand side of (1.38) with respect to α_i.

If $(p_i, Cp_i) = 0$, then (1.38) does not contain a quadratic term and if $(p_i, f'(x_0)) \neq 0$ then the lower bound with respect to α_i is $-\infty$. If $(p_i, f'(x_0)) = 0$, then α_i may be arbitrarily chosen, since f is actually independent of α_i.

Let x_0 denote an arbitrary point. Then we deduce that a necessary and sufficient condition for the function $f(x)$ of (1.36) to be bounded below is that the equation

$$(p_i, f'(x)) = 0$$

is satisfied for all x and for all i such that $(p_i, Cp_i) = 0$.

If $(p_i, Cp_i) \neq 0$, then minimizing (1.38) with respect to α_i gives

$$\alpha_i = -(p_i, f'(x_0))/(p_i, Cp_i). \tag{1.39}$$

Solely from the above, it follows that if C is a positive-definite matrix, then either the quadratic function f is not bounded below or it attains a minimum at some point.

If a minimum is attained, this minimum is of the form

$$x_* = x_0 + \sum_{i=1}^{n} \alpha_i p_i,$$

where the α_i are given by (1.39), if $(p_i, Cp_i) \neq 0$, and are arbitrary otherwise. To be definite, in this case, we set $\alpha_i = 0$.

We next suppose that

$$x_{i+1} = x_i + \alpha_{i+1} p_{i+1}, \quad i = 0, 1, \ldots, n-1.$$

We note that

$$\begin{aligned} f'(x_i) &= f'(x_i) - f'(x_{i-1}) + f'(x_{i-1}) - f'(x_{i-2}) \\ &\quad + \ldots + f'(x_1) - f'(x_0) + f'(x_0) \\ &= \alpha_i C p_i + \alpha_{i-1} C p_{i-1} + \ldots + \alpha_1 C p_1 + f'(x_0). \end{aligned}$$

Thus,

$$(p_{i+1}, f'(x_i)) = (p_{i+1}, f'(x_0)).$$

It now follows that the points x_i, $i = 1, \ldots, n$ may be obtained using the recurrence relations

$$\begin{aligned} x_{i+1} &= x_i + \alpha_{i+1} p_{i+1} \\ \alpha_{i+1} &= -(p_{i+1}, f'(x_i))/(p_{i+1}, Cp_{i+1}), \quad i = 0, 1, \ldots, n-1 \end{aligned}$$

and $x_n = x_*$ is a solution of the problem.

If, during the process, it turns out that $(p_i, Cp_i) = 0$ but $(p_i f'(x_{i-1})) \neq 0$, then it follows directly that the function $f(x)$ is not bounded below.

It remains to find methods for constructing conjugate directions. This question may be solved by many techniques, which give rise to various algorithms. In this respect, the reader may refer to the literature listed in the References. Here, we shall only describe one algorithm which is the simplest and most convenient for minimizing quadratic functions.

1.3.2 Conjugate Gradient Algorithm

Here, we describe a form of the conjugate gradient algorithm which is convenient for the minimization of a quadratic function.

An arbitrary initial point x_0 is chosen. All other calculations for $i = 0, 1, \ldots, n-1$ are carried out using the recurrence relations:

$$\begin{aligned}
\beta_{i+1} &= \begin{cases} 0 & i = 0 \\ \frac{(f'(x_i), f'(x))}{(f'(x_{i-1}), f'(x_{i-1}))} & i > 0 \end{cases} \\
p_{i+1} &= \begin{cases} f'(x_i) & i = 0 \\ f'(x_i) + \beta_{i+1} p_i & i > 0 \end{cases} \\
\alpha_{i+1} &= -\frac{(p_{i+1}, f'(x_i))}{(p_{i+1}, C p_{i+1})} \\
x_{i+1} &= x_i + \alpha_{i+1} p_{i+1}.
\end{aligned}$$

Remark.

1. As we saw in Section 1.3.1, we only need to know the vector p_{i+1} after the point x_i has been constructed. The above process constructs the conjugate vectors as they are needed.

2. A condition for the process to stop is that the gradient $f'(x_i)$ should become zero. This takes place for $i = n$, but may occur earlier if the point x_0 is favourably chosen or if some of the eigenvalues of the matrix C are zero.

3. It is not known a priori whether the function $f(x)$ is bounded below. Thus, when calculating α_{i+1}, there may arise a situation in which

$$(p_{i+1}, C p_{i+1}) = 0, \quad (p_{i+1}, f'(x_i)) \neq 0.$$

   The occurrence of such a situation immediately implies that the function $f(x)$ is not bounded below.

4. Calculations show that when the matrix C is ill-conditioned (i.e. when it is nonsingular, but the ratio of the greatest eigenvalue to the least is very large), this process is subject to computational error. Thus, if $f'(x_n)$ is not sufficiently close to zero, the algorithm will have to be applied again taking the point x_n as the starting point; in other words, the algorithm will have to be used as one step in a more general iterative process.

1.3.3 Existence of a Solution

We shall now turn to the general problem of quadratic programming. This is the problem of minimizing the quadratic function

$$f(x) = 1/2(x, Cx) + (d, x)$$

subject to the constraints:

$$\begin{aligned} (a_i, x) &\leq \alpha_i, \quad i \in I^- \\ (a_i, x) &= \alpha_i, \quad i \in I^0 \end{aligned} \tag{1.40}$$

where the matrix C is positive definite and I^- and I^0 are finite sets of indices. At this point, we are interested in whether the quadratic programming problem is solvable.

We note that, in the constraints (1.40), the equalities may be eliminated if some of the unknowns in these equations are expressed in terms of the others and substituted in the remaining inequalities and in the expression for the function f. This decreases the number of unknowns and leaves only constraints in the form of inequalities. Thus, at least for current purposes, we may assume that the problem has the form

$$\min\{f(x) | (a_i, x) \leq \alpha_i, \quad i = 1, \ldots, m\}. \tag{1.41}$$

Theorem 1.15 *In the quadratic programming program, either a minimum is attained at some point or the function f is not bounded below.*

Proof. We shall use induction on the number of inequalities m. For $m = 0$, there are no constraints and the theorem follows from Section 1.3.1. Let us assume that it is true for m inequalities and show that it is also true for $m + 1$ inequalities.

We introduce the notation:

$$\begin{aligned} v_{m+1} &= \inf_x\{f(x) | (a_i, x) \leq \alpha_i, \quad i = 1, \ldots, m+1\} \\ v_m &= \inf_x\{f(x) | (a_i, x) \leq \alpha_i \quad i = 1, \ldots, m\} \\ W &= \inf_x\{f(x) | (a_i, x) \leq \alpha_i, \quad i = 1, \ldots, m, \quad (a_{m+1}, x) = \alpha_{m+1}\} \end{aligned}$$

The problems on the right-hand sides of these formula are denoted by P_{m+1}, P_m and P_0, respectively. Clearly, $v_m \leq v_{m+1} \leq W$. Furthermore, in problem P_0, either a minimum is attained or $W = -\infty$, since this problem only involves m inequalities.

If $W = -\infty$, then $v_{m+1} = -\infty$ and the theorem is proved. Thus, we shall assume that W is finite with minimum y. Then, if $v_{m+1} = -\infty$, the theorem is also proved. Thus, we must assume that v_{m+1} is finite and prove that there exists a point z satisfying the constraints of problem P_{m+1} with $f(z) = v_{m+1}$.

Two cases are then possible.

a) v_m is finite and a minimum is attained at the point $\overline{z}$. If

$$(a_{m+1}, \overline{z}) \leq \alpha_{m+1}$$

then a minimum of P_{m+1} is attained at $\overline{z}$ and everything is true. Thus, we suppose that

$$(a_{m+1}, \overline{z}) > \alpha_{m+1} \tag{1.42}$$

We introduce an arbitrary point x satisfying the constraints of problem P_{m+1} and associate it with the point

$$\overline{x} = x + t(\overline{z} - x), \quad 0 \leq t \leq 1$$

where t is chosen such that $(a_{m+1}, \overline{x}) = \alpha_{m+1}$. Such a t exists since

$$(a_{m+1}, x) \leq \alpha_{m+1}$$

and (1.42) is satisfied.

Clearly, $\overline{x}$ satisfies the constraints of problem P_0 (in fact, x satisfies P_{m+1} and $\overline{z}$ satisfies P_m), and since f is a convex function, we have

$$f(\overline{x}) \leq (1-t)f(x) + tf(\overline{z}) \leq f(x)$$

because $f(\overline{z}) = v_m \leq v_{m+1} \leq f(x)$. Thus, if y is a solution of problem P_0, then y satisfies the constraints of problem P_{m+1} and $f(y) \leq f(\overline{x}) \leq f(x)$; in other words, y is a solution of problem P_{m+1} and a minimum of P_{m+1} is attained.

b) $v_m = -\infty$. Suppose that x_k, $k = 1, 2, \ldots$, is a minimizing sequence for problem P_m, i.e. $f(x_k) \to -\infty$. If arbitrarily many points of the sequence x_k satisfy the constraints of problem P_{m+1}, then $v_{m+1} = -\infty$. Thus, we shall assume that, for k sufficiently large, the inequality

$$(a_{m+1}, x_k) > \alpha_{m+1} \tag{1.43}$$

is satisfied.

If x is a feasible point for problem P_{m+1}, we then choose a larger k which satisfies (1.43) and $f(x) \geq f(x_k)$ and associate the point x with the point

$$\overline{x} = x + t(x_k - x) \quad 0 \leq t \leq 1,$$

such that

$$(a_{m+1}, \overline{x}) = \alpha_{m+1}.$$

Then $\overline{x}$ is again a feasible point for problem P_0 and

$$f(\overline{x}) \leq (1-t)f(x) + tf(x_k) \leq f(x).$$

Whence, as before, we deduce that the solution y of problem P_0 is also a solution of problem P_{m+1}

The proof of the theorem is complete □

1.3.4 Necessary Conditions for an Extremum and the Dual Problem

Using the results of Section 1.2, it is easy to write down necessary conditions for an extremum in a quadratic programming problem. Indeed, application of Theorem 1.3 and the remark following it lead to the following result:

Theorem 1.16 *Suppose that x_0 is a minimum of the quadratic programming problem with constraints (1.40). Then, there exist numbers u^i, $i \in I = I^- \cup I^0$, such that*

$$\begin{aligned} f'(x_0) + \sum_{i\in I} u^i a_i &= 0 \\ u^i \geq 0, \quad u^i[(a_i, x_0) - \alpha_i] &= 0, \quad i \in I. \end{aligned}$$

To form the dual problem, we write down the Lagrange function

$$\begin{aligned} L(x,u) &= 1/2(x, Cx) + (x, d) + \sum_{i\in I} u^i[(a_i, x) - \alpha_i] \\ &= 1/2(x, Cx) + (x, d + \sum_{i\in I} u^i a_i) - \sum_{i\in I} u^i \alpha_i \\ &= 1/2(x, Cx) + (x, d + A^* u) - (b, u) \end{aligned}$$

where b is the vector with components α_i, $i \in I$, u is the vector with components u^i, $i \in I$ and A is the matrix with rows the vectors a_i^*, $i \in I$. We recall that a_i is a column vector, so that a_i^* is a row vector. In order to construct the dual problem, according to Section 1.2.4, we need to calculate

$$\varphi(u) = \inf_x L(x, u).$$

We assume that the matrix C is nonsingular. Setting the derivatives of L with respect to x to zero, we obtain

$$\begin{aligned} L'_x(x, u) &= Cx + d + A^* u = 0 \\ x &= -C^{-1}[d + A^* u]. \end{aligned} \tag{1.44}$$

Substituting this in the expression for $L(x, u)$, we have

$$\begin{aligned} \varphi(u) &= 1/2(C^{-1}[d + A^*u], [d + A^*u]) - (C^{-1}[d + A^*u], d + A^*u) - (b, u) \\ &= -1/2(C^{-1}[d + A^*u], d + A^*u) - (b, u). \end{aligned}$$

Thus, if C is a positive-definite nonsingular matrix, then, according to Section 1.2.4, the dual quadratic programming problem involves maximizing the quadratic function

$$\varphi(u) = -1/2(C^{-1}[d + A^*u], d + A^*u) - (b, u) \tag{1.45}$$

with the constraints $u^i \geq 0$, $i \in I^-$.

Here, it is understood that the constraints of (1.40) corresponding to $i \in I^0$ may be written as two inequalities, which, according to the remark following Theorem 1.3, leads to an indeterminate sign of the corresponding factor u^i.

As shown in Section 1.3.3, the minimum in the quadratic programming problem is always attained and thus Theorem 1.9 applies. It follows from this theorem that there exists a Kuhn–Tucker vector which is a solution of the dual problem.

If now a solution u of problem (1.45) is found, then substituting it in (1.44), we obtain a solution of the original problem. Indeed, since C is positive definite and nonsingular, it is easy to show that the original problem has a unique solution x_0. According to Theorem 1.6, the point x_0 must be a minimum of $L(x,u)$, if u is a Kuhn–Tucker vector, i.e. a solution of problem (1.45). But this minimum is unique and is attained by the expression given in formula (1.44).

Theorem 1.17 *Suppose that C is a positive-definite, nonsingular matrix. Then, the problem of maximizing the function (1.45) subject to the constraints $u^i \geq 0$ is the dual quadratic programming problem. If it has solution u, then $x_0 = -C^{-1}[d + A^*u]$ is the solution of the original problem.*

We apply this theorem to the special case when there are no inequality-type constraints, i.e. when $I^- = \emptyset$. Then, using the previous notation, the constraints of (1.40) corresponding to $i \in I^0$ may be written in the form

$$Ax - b = 0 \tag{1.46}$$

According to Theorem 1.17, a solution of the problem of minimizing the function $f(x)$ given by formula (1.36) subject to the constraints (1.46) has the form (1.44), where there are no constraints on the sign of u. Substituting formula (1.44) in (1.46), we obtain

$$AC^{-1}[d + A^*u] + b = 0. \tag{1.47}$$

We then note that if the rows of the matrix A (i.e. $a_i^*, i \in I^0$) are linearly independent, then the matrix $AC^{-1}A^*$ is invertible. In fact, this matrix is not invertible if it transforms some nonzero vector v to zero:

$$AC^{-1}A^*v = 0.$$

But then

$$v^*AC^{-1}A^*v = (C^{-1}(A^*v), A^*v) = 0$$

which is only possible if $A^*v = 0$, i.e. if the rows of the matrix A, being columns of the matrix A^*, are linearly dependent, which contradicts the assumption. Thus, with the assumptions of (1.47) and (1.44), we obtain

$$u = -(AC^{-1}A^*)^{-1}[b + AC^{-1}d] \tag{1.48}$$

$$x_0 = -C^{-1}[d + A^*u]. \tag{1.49}$$

Thus, we have proved the following theorem.

Theorem 1.18 *Suppose that, in the minimization problem (1.36) subject to the constraints (1.46) the matrix C is nonsingular and the rows of the matrix A are linearly independent. Then the solution of the direct and the dual problem is given by formulae (1.48) and (1.49).*

1.3.5 Application. Projection onto a Subspace

The problem of projecting a vector onto a subspace is a problem of quadratic programming. By definition, a *projection* of a vector y onto a subspace $M = \{x|Ax = 0\}$ is a point $x_0 \in M$, such that the distance between x_0 and y is a minimum for the Euclidean norm. Thus, construction of a projection reduces to finding a solution of the problem

$$\min_x\{1/2\|x - y\|^2 | Ax = 0\}.$$

Using the fact that

$$1/2\|x - y\|^2 = 1/2(x, x) - (x, y) + 1/2(y, y),$$

the projection x_0 may be obtained by applying Theorem 1.18, where in formulae (1.48) and (1.49), we must set $C = I$ (the unit matrix), $b = 0$ and $d = -y$.

Thus,

$$u = (AA^*)^{-1}Ay, \quad x_0 = y - A^*(AA^*)^{-1}Ay.$$

We denote

$$P = A^*(AA^*)^{-1}A. \tag{1.50}$$

Theorem 1.19 *Suppose that $M = \{x|Ax = 0\}$ and that the rows of the matrix A are linearly independent. Then the projection of the vector y onto the subspace M is given by the formula*

$$x = (I - P)y.$$

We now mention a number of properties of the matrix P, which follow directly from its definition (1.50):

a) $PA^* = A^*$;

b) $P^2 = P, \quad P(I - P) = 0, \quad P^* = P$;

c) $A(I - P) = 0$.

From property b), we obtain that, for an arbitrary vector y,

$$\begin{aligned} y &= Py + (I - P)y \\ (Py, (I - P)y) &= (y, P(I - P)y) = 0. \end{aligned}$$

Thus, each vector y may be divided into the sum of two orthogonal vectors, one of which is the projection of y onto M and the other of which belongs to the orthogonal complement of M, the space $M^{\perp}$. In fact, for any $x \in M$, by virtue of b),

$$(Py, x) = (y, Px) = 0,$$

which implies that $Py \in M^\perp$.

If now y is an arbitrary vector and $x \in M^\perp$, then we have the chain of equations

$$\begin{aligned} \|y - x\|^2 &= \|Py - x + (I - P)y\|^2 \\ &= \|Py - x\|^2 + 2(Py - x, (I - P)y) + \|(I - P)y\|^2. \end{aligned}$$

But, from property b) and the fact that $x \in M^\perp$ and $(I - P)y \in M$, we have

$$(Py, (I - P)y) = 0, \quad (x, (I - P)y) = 0;$$

whence,

$$\|y - x\|^2 = \|Py - x\|^2 + \|(I - P)y\|^2.$$

Therefore, it is clear that the distance between y and $x \in M^\perp$ is a minimum when $x = Py$. Thus, P is the projection operator onto the subspace $M^\perp$.

1.3.6 Algorithm for the Quadratic Programming Problem

We shall now study the possibilities for practical resolution of the quadratic programming problem. Here, we shall describe one of the many effective algorithms. We chose this method for three reasons:

1. It is finite, in other words, it converges after a finite number of steps;

2. It is a generalization of the simplex method of linear programming; thus, a whole arsenal of techniques used to solve linear programming problems, including, in particular, techniques for working with sparse matrices, may be used here.

3. It may be extended to the problem of minimizing arbitrary functions with linear constraints.

We consider the problem

$$\min\{1/2(x, Cx) + (d, x) | Ax = b, \ \ x \geq 0\}, \tag{1.51}$$

where C is a positive-definite matrix, A is an $m \times n$ matrix and $b \in \mathbb{R}^n$. Problem (1.51) is called a *quadratic programming problem in standard form.*

Any problem with general constraints (1.40) may be brought to the standard form using the following techniques.

If in (1.40), $I \neq \emptyset$, then for each $i \in I^-$, we introduce an additional variable z^i and the inequality

$$(a_i, x) \leq \alpha_i$$

is replaced by the equation

$$(a_i, x) + z^i = \alpha_i, \quad z^i \geq 0.$$

Thus, in the new variables, all the constraints are equality-type constraints and inequality-type constraints have a simple form $x^i \geq 0$ or $z^i \geq 0$.

If, after these transformations, some component x^j is not required to be positive, then we may replace x^j by two new variables x^{j+} and x^{j-} using the equation

$$x^j = x^{j+} - x^{j-}, \quad x^{j+} \geq 0, \quad x^{j-} \geq 0,$$

after which the problem takes on the standard form. In practice, we do not actually need to carry out this substitution, and must simply calculate

$$\begin{aligned} x^j &= x^{j+}, \quad x^{j-} = 0 \text{ if } x^j \geq 0 \\ x^j &= -x^{j-}, \quad x^{j+} = 0 \text{ if } x^j \leq 0. \end{aligned}$$

Thus, without loss of generality, it suffices to consider only the problem in standard form.

We now introduce some notation which will apply only in this section.

We denote the columns of the matrix A by A_j. Then

$$Ax = \sum_{j=1}^{n} A_j x^j$$

Suppose that J is some subset of indices of $\{1, 2, \ldots, n\}$. We shall use $x|_J$ to denote the vector with components $x^j, j \in J$, where the components are ordered in increasing order of their indices. In the same way, $A|_J$ is the $m \times |J|$ matrix, where $|J|$ is the number of elements in the set J, with columns A_j, $j \in J$. We also use the notation $f'|_J(x)$ for the row vector with components $\partial f / \partial x^j$, $j \in J$. It is also convenient to use $\overline{J}$ to denote the complement of J in $\{1, 2, \ldots, n\}$.

With this notation, for example, the following equation holds

$$A|_J x|_J + A|_{\overline{J}} x|_{\overline{J}} = Ax.$$

We shall say that the vectors A_j, $j \in \overline{J}$ form a *basis* if $|\overline{J}| = m$ and the A_j are linearly independent. The corresponding variables x^j are called *basis variables*. Thus, if A_j, $j \in \overline{J}$ is a basis, then the matrix

$$B_{\overline{J}} = (A|_{\overline{J}})^{-1}$$

is defined. For a given basis, the basis variables are easily expressed in terms of the others via the equation

$$Ax = A|_{\overline{J}} x|_{\overline{J}} + A|_J x|_J = b$$

using the formula

$$x|_{\overline{J}} = B_{\overline{J}}b - B_{\overline{J}}A|_J x|_J. \tag{1.52}$$

If we denote the matrix with components $\partial x^j/\partial x^k$, $j \in \overline{J}, k \in J$ by

$$\partial x|_{\overline{J}}/\partial x|_J,$$

then it follows from (1.52) that

$$\partial x|_{\overline{J}}/\partial x|_J = -B_{\overline{J}}A|_J.$$

Suppose that the basis variables are expressed in terms of non-basis variables according to (1.52) and substituted in $f(x)$. We denote the corresponding function by $f|_J(x|_J)$. Then, according to well-known formulae for differentiation,

$$(f|_J)' = f'|_J + f'|_{\overline{J}}\frac{\partial x|_{\overline{J}}}{\partial x|_J}$$

or

$$(f|_J)' = f'|_J - f'|_{\overline{J}}B_J A|_J. \tag{1.53}$$

We note that here, $(f|_J)'$ is a row vector consisting of the derivatives of the composite function $f|_J$ of the variables x^j, $j \in J$, while at the same time, for example, $f'|_J$ is a row vector consisting of the derivatives $\partial f/\partial x^j$, $j \in J$ of the original function.

We now describe the main idea of the algorithm. Suppose that an initial point x is chosen, satisfying the constraints of the problem. We then choose a basis $\overline{J}$. The remaining, indices, (i.e. the set J) are divided into active and passive indices, using a criterion which we shall discuss separately: J_{a} and J_{p}, with $J_{\mathrm{a}} \cup J_{\mathrm{p}} = J$. This division depends on the point x. We simply note that we always have

$$x^j = 0,\ j \in J_{\mathrm{p}}. \tag{1.54}$$

Constraining the passive variables to zero, we change the active variables, by expressing the basis variables in terms of these, according to formula (1.52). Since the basis variables are expressed as linear combinations of the others, after substitution in $f(x)$, we again obtain a quadratic function of the active variables. We shall minimize this function with respect to the active variables using the conjugate gradient method. However, because of the constraints on the signs of the variables, we need to monitor the progress of the conjugate gradient process. If a step of the process leads to one of the variables becoming negative, then the length of the step is limited so that this variable is transformed to zero exactly. After this transformation to zero the variable becomes passive. When an active variable is transformed to zero the process of the conjugate gradient method is further applied to the remaining active variables. If a basis variable is transformed to zero, then the basis is changed, the variable transformed to zero becomes passive and one of the active variables is introduced into the basis.

As can be seen from the above, this process cannot continue indefinitely, since the number of passive variables increases all the time and the number of active variables decreases. Thus, at any instant the number of active variables is fixed and the conjugate gradient method finds a minimum of the quadratic function with respect to the active variables. The point obtained is selected as the initial point and the process is continued.

That the algorithm is finite is a result of the following factors:

1. Since the initial point satisfies the constraints of the problem and the basis variables are expressed in terms of formula (1.52), all the constraints of the problem are satisfied during the process.

2. The whole process consists of major steps similar to those described above and each step ends after a finite number of operations to find the minimum with respect to the active variables for fixed passive variables.

3. The number of possible choices of groups of passive variables is finite.

4. The process is monotonic, i.e. the function $f(x)$ is strictly decreasing all the time; this excludes the possibility of repetition of groups of passive variables, constrained to zero, for which a minimum is found.

5. After each major iteration step the process begins again with the point obtained; if this is not a solution of the problem there is a nonzero decrease in the function f.

Thus, the algorithm should be constructed so that the factors listed above are complied with.

For the algorithm to work successfully, the problem must satisfy a nondegeneracy requirement similar to the analogous requirement in linear programming.

Nondegeneracy condition Suppose that x satisfies the constraints $Ax = b$, $x \geq 0$ and that $\overline{J}(x) = \{j|x^j > 0\}$. Then there exists a set $\overline{J} \subseteq \overline{J}(x)$, $|\overline{J}| = m$, such that the vectors A_j, $j \in \overline{J}$ are linearly independent.

Assuming in what follows that this condition is satisfied, we shall formulate the algorithm. We describe one major step of the algorithm.

0. Let x_0 be a point satisfying the constraints $Ax = b$, $x \geq 0$.

   We introduce a basis $\overline{J}^0$ from the subset of columns $\overline{J}(x_0)$. This is possible because of the nondegeneracy condition.

1. We set J^0 equal to the complement of $\overline{J}^0$ in $\{1, \ldots, n\}$. We compute $B_{\overline{J}^0}$ and the vector

$$(f|_{J^0})' = f'|_{J^0} - f'|_{\overline{J}^0} B_{\overline{J}^0} A|_{J^0}$$

   corresponding to the point x_0. If for $j \in J^0$

$$\begin{aligned} (f|_{J^0})'|_j &= 0 \text{ for } x_0^j > 0 \\ (f|_{J^0})'|_j &\geq 0 \text{ for } x_0^j = 0 \end{aligned}$$

   then x_0 is a solution of the problem.

2. Suppose that

$$\begin{aligned} J_{\mathrm{a}}^0 &= \{j | x_0^j > 0,\ (f|_{\overline{J}^0})'|_j \neq 0 \ \text{or}\ (f|_{J^0})'|_j < 0,\ x_0^j = 0, j \in J^0\} \\ J_{\mathrm{p}}^0 &= J^0 \backslash J_{\mathrm{a}}^0. \end{aligned}$$

3. For $k = 0, 1, \ldots$ we apply the conjugate gradient method to minimize the function $f|_{J^k}$ obtained from f by replacing $x|_{\overline{J}^k}$ in expression (1.52) by

$$x|_{\overline{J}^k} = B_{\overline{J}^k} b - B_{\overline{J}^k} A|_{J^k} x|_{J^k}.$$

The gradient of this function is computed from formula (1.53). Here, the minimization is only carried out with respect to the arguments x^j, $j \in J_{\mathrm{a}}^k$.

The function $f|_{J^k}$ is quadratic in $x|_{J^k}$, but the matrix C_{J^k} determining the quadratic part is not described explicitly. Thus, the following method is used to compute the step α in the direction $p|_{J_{\mathrm{a}}^k}$ from the point x.

The vector $p|_{J_{\mathrm{a}}^k}$ is determined from the formulae of the conjugate gradient method. We also assume that

$$p|_{J_{\mathrm{p}}^k} = 0,\ \ p|_{\overline{J}^k} = -B_{\overline{J}^k} A|_{J^k} p|_{J^k}.$$

According to the formulae of the conjugate gradient method

$$\alpha = -(p|_{J^k}, (f|_{J^k})'(x))/(p|_{J^k}, C_{J^k} p|_{J^k}). \tag{1.55}$$

The numerator in (1.55) is computed from the previous formulae for $(f|_{J^k})'$, and

$$C_{J^k} p|_{J^k} = (f|_{J^k})'(x+p) - (f|_{J^k})'(x),$$

since for any quadratic function f

$$f'(x) = Cx + d,\ \ f'(x+p) - f'(x) = Cp.$$

A number of other features are taken into account. If the denominator in (1.55) is equal to zero and the numerator is also equal to zero, the application of the conjugate gradient method terminates, the current point is chosen as the initial point and stage 0 is returned to.

If the numerator is less than zero (only this case may arise, by the construction of the conjugate gradient method), then we set $\alpha = +\infty$.

We compute

$$\overline{\alpha} = \min\{-x^j/p^j | j \in J_{\mathrm{a}}^k \cup \overline{J}^k,\ p^j < 0\}.$$

If $\{j \in J_{\mathrm{a}}^k \cup \overline{J}^k | p^j < 0\} = \emptyset$, then $\overline{\alpha} = +\infty$.

If $\alpha < \overline{\alpha}$ then we construct a new point $x + \alpha p$ and the process of minimizing $f|_{J^k}$ with respect to x^j, $j \in J_{\mathrm{a}}^k$ continues.

If $\alpha \geq \overline{\alpha}$, then we construct a new point $\overline{x} = x + \overline{\alpha}p$.

In the case where $\overline{\alpha} = +\infty$, the function f is not bounded below and the operation of the algorithm terminates. If $\overline{\alpha} < +\infty$, then the sets J_{a}^k, J_{p}^k and $\overline{J}^k$ are reconstructed. Namely, we suppose that $j \in J_{\mathrm{a}}^k \cup \overline{J}^k$ is the index such that $\overline{\alpha} = -x^j/p^j$. Then we set

$$J_{\mathrm{p}}^{k+1} = J_{\mathrm{p}}^k \cup \{j\}, \quad J_{\mathrm{a}}^{k+1} = J_{\mathrm{a}}^k \backslash \{j\}, \quad \overline{J}^{k+1} = \overline{J}^k$$

if $j \in J_{\mathrm{a}}^k$. But if $j \in \overline{J}^k$, then

$$\overline{J}^{k+1} = \{\overline{J}^k \backslash \{j\}\} \cup \{l\}, \quad J_{\mathrm{a}}^{k+1} = J_{\mathrm{a}}^k \backslash \{l\},$$

where $l \in \overline{J}(x+\alpha p) \backslash \overline{J}^k$ is the index for which $x^l + \overline{\alpha} p^l > 0$, and the vectors A_j, $j \in \overline{J}^{k+1}$ are linearly independent. After this, stage 3 is returned to.

After a finite number of operations for some k (in fact, the set J_{p}^k is increasing all the time) the current point x is a minimum of $f|_{J^k}$ with respect to x^j, $j \in J_{\mathrm{a}}^k$. This point becomes the initial point and stage 0 is returned to.

4. The process ends in stage 2 if $J_{\mathrm{a}}^0 = \emptyset$.

The basis for the finite convergence of the algorithm has already been described in general terms. We need only mention a few details here.

If the set J_{a}^0 is nonempty, then at the first step of the application of the conjugate gradient method the components of the vector p along which the displacement from the point x_0 occurs coincide, for $j \in J_{\mathrm{a}}^0$, with the derivatives of the function $f|_{J^0}$ taken with a minus sign. These derivatives are nonzero if $x_0^j > 0$ and are strictly negative if $x_0^j = 0$. Thus, the first step always leads to a successful nonzero displacement and a strict decrease in the goal function.

Let us assume that the stopping criterion of stage 1 is satisfied. We take $u \in \mathbb{R}^m$ and $v \in \mathbb{R}^n$, such that

$$u^* = -f'|_{\overline{J}^0} B_{\overline{J}^0} \tag{1.56}$$

$$v^*|_{J^0} = (f|_{J^0})' \tag{1.57}$$

$$v^*|_{\overline{J}^0} = 0. \tag{1.58}$$

Then, by virtue of (1.53) and (1.56)–(1.58),

$$\begin{aligned} f'|_{J^0} + u^* A|_{J^0} &= (f|_{J^0})' = v^*|_{J^0} \\ f'|_{\overline{J}^0} + u^* A|_{\overline{J}^0} &= 0 = v^*|_{\overline{J}^0}, \end{aligned}$$

where we have used the fact that $B_{\overline{J}} = (A|_J)^{-1}$. Using the previous notation these two equations may be combined into one:

$$f' + u^* A = v^*.$$

Next, we note that, from the stopping criterion in stage 1 of the algorithm,

$$v \geq 0, \quad v^j x_0^j = 0.$$

Comparing this with Theorem 1.4, we obtain that x_0 is a solution of the quadratic programming problem. This completes the proof that the algorithm converges.

We note that, we have simultaneously obtained formulae (1.56)–(1.58) for computing the Kuhn–Tucker coefficients.

We also note the fact that the equation $J_{\mathrm{a}}^k = \emptyset$, $k \geq 1$ only shows that the minimization process based on the conjugate gradient method terminates. The equation $J_{\mathrm{a}}^k = \emptyset$ only shows that, for fixed passive variables, it is impossible to achieve a further decrease in the value of the function.

Furthermore, by construction, the basis variables are always strictly positive. This, together with the fact that it is possible to reconstruct a basis when one of these elements is transformed to zero, is guaranteed by the nondegeneracy condition.

1.3.7 Computational Aspects

In the preceding discussions, we left aside a number of important practical questions, which are crucial to the effectiveness of the given algorithm. Some of these questions have already been successfully resolved in linear programming and will only be discussed briefly here. A more detailed consideration of these may be found in the literature relating to this subsection.

To begin work with the algorithm, we need a point satisfying the constraints of the algorithm, $Ax = b$, $x \geq 0$. If no such point is known from a priori information any standard linear programming program may be used. Such programs are very common and we shall not consider this question further. We shall now concentrate on the computation of the matrices $B_{\bar{J}}$ and $B_{\bar{J}}A_{\bar{J}}$ needed during the operation of the algorithm.

Suppose we solve the $m \times n$ system of equations

$$Ax = b \tag{1.59}$$

where the matrix A has m linearly independent columns. We denote the rows of the matrix A by a_i, the columns by A_j and the elements by a_{ij}.

The standard method for solving the system (1.59) involves using the ith equation, if $a_{ij} \neq 0$ to express x^j in terms of the remaining variables and then substituting this in the remaining equations.

This method of eliminating unknowns is known as the *Gauss–Jordan elimination method.* The next unknown is eliminated from the transformed system etc.

We shall describe this method in a form which is convenient for what follows. It is well-known to be equivalent to the following transformation: the ith row of the matrix should be multiplied by the factor t_k and added to the kth row, where t_k is chosen such that the transformed element a_{kj} is equal to zero if $k \neq i$ or to one if $k = i$.

If the transformed matrix is denoted by $\overline{A}$, then

$$\begin{aligned} \overline{a}_k &= a_k + t_k a_i, \quad t_k = -a_{kj}/a_{ij}, \quad k \neq i \\ \overline{a}_i &= t_i a_i, \quad t_i = 1/a_{ij}. \end{aligned} \tag{1.60}$$

It follows from these formulae, that if $a_{il} = 0$, then $\overline{a}_{kl} = a_{kl}$ for all of column l.

We introduce the $m \times m$ matrix

$$T_{ij} = \begin{pmatrix} 1 & 0 & \dots & 0 & t_1 & 0 & \dots & & 0 \\ \dots & \dots & \dots & \dots & \dots & \dots & \dots & \dots & \dots \\ 0 & & \dots & 1 & t_{i-1} & 0 & \dots & & 0 \\ 0 & & \dots & 0 & t_i & 0 & \dots & & 0 \\ 0 & & \dots & 0 & t_{i+1} & 1 & \dots & & 0 \\ \dots & \dots & \dots & \dots & \dots & \dots & \dots & \dots & \dots \\ 0 & & \dots & 0 & t_m & 0 & \dots & 0 & 1 \end{pmatrix} \tag{1.61}$$

This clearly has determinant $t_i \neq 0$ and is therefore a nonsingular matrix. Then, it is easy to see that the above transformation leads to a new system

$$\overline{A}x = \overline{b}, \quad \overline{A} = T_{ij}A, \quad \overline{b} = T_{ij}b.$$

Here, the variable x^j is only involved in the ith equation with coefficient 1:

$$\overline{a}_{kj} = 0, \quad k \neq i, \quad \overline{a}_{ij} = 1.$$

Alternatively, this may be written in the form

$$\overline{A}_j = e_i$$

where e_i is the vector with components all equal to zero except for the ith which is equal to 1.

We next proceed as follows. We set $i = 1$ and find $a_{1j_1} \neq 0$. We denote $j(1) = j_1$. We set

$$T^{(1)} = T_{1j_1}, \quad A^{(1)} = T^{(1)}A, \quad b^{(1)} = T^{(1)}b.$$

Analogously, setting $i = 2$, we find a nonzero element $a^{(1)}_{2j_2} \neq 0$ amongst the elements $a^{(1)}_{2j}$. Such an element can certainly be found since otherwise the whole of the second row of $A^{(1)}$ would be zero and the rank of $A^{(1)}$ would be less than m. But, since the matrix A has rank m and the matrix $T^{(1)}$ is nonsingular, the rank of $A^{(1)}$ is also m. Moreover, $j_2 \neq j_1$, since $a_{kj_1} = 0$, $k \neq 1$, by the construction of $A^{(1)}$, so that $a_{2j_1} = 0$.

We set

$$j(2) = j_2, \quad T^{(2)} = T_{2j_2}, \quad A^{(2)} = T^{(2)}A^{(1)} = T^{(2)}T^{(1)}A, \quad b^{(2)} = T^{(2)}T^{(1)}b,$$

where the elements of the matrix T_{2j_2} are constructed from the elements of the matrix $A^{(1)}$. We note here that since $a_{2j}^{(1)} = 0$, then $A_{j_1}^{(2)} = A_{j_1}^{(1)} = e_1$.

Continuing analogously, after m steps, we obtain the matrices

$$\begin{aligned} A^{(m)} &= T^{(m)} \dots T^{(1)} \\ A &= T^{(m)} A^{(m-1)}, \quad b^{(m)} = T^{(m)} \dots T^{(1)}, \quad b = T^{(m)} b^{(m-1)} \end{aligned} \tag{1.62}$$

and a one-to-one mapping $j(i) = j_i$. Because this mapping is one-to-one, the inverse mapping $i(j)$ is defined for $j \in \{j_1, j_2, \dots j_m\}$ and is given by $i(j_k) = k$. By construction, the column with index j_k coincides with the kth unit vector e_k:

$$A_{jk}^{(m)} = e_k \tag{1.63}$$

We denote $\overline{J} = \{j_1, \dots, j_m\}$. The columns $A_j^{(m)}$, $j \in \overline{J}$ form a basis for A^m since they coincide with the different unit vectors. Since the transformations involved were nonsingular, the columns A_j, $j \in \overline{J}$ of the original matrix are linearly independent and form a basis. Generally, the property of linear dependence or independence is preserved for columns of the original and transformed matrices.

We then form the matrix $\overline{A}_J = [A_{j_1} \dots A_{j_m}]$. From formulae (1.62) and (1.63), it follows that

$$\overline{B}_{\overline{J}} \overline{A}_{\overline{J}} = I_m, \quad \overline{B}_{\overline{J}} = T^{(m)} \dots T^{(1)} \tag{1.64}$$

where I_m is the unit $m \times m$ matrix.

Thus, $\overline{B}_{\overline{J}}$ is the inverse of the matrix $\overline{A}_{\overline{J}}$. However, $\overline{B}_{\overline{J}}$ is not the same as $B_{\overline{J}} = (A|_{\overline{J}})^{-1}$ described in Section 1.3.6. In fact, in the matrix $A|_{\overline{J}}$, the columns A_j are ordered in the natural initial ordering. However, it is easy to see that this does not cause complications. Clearly, rearrangement of the columns of A is equivalent to a simple rearrangement of the variables and does not affect the essence of the problem. Thus, if the basis $\overline{J}$ is chosen and the function $j(i)$, $i = 1, \dots, m$ is known, then in all the computations needed in the execution of the algorithm of Section 1.3.6 involving this basis, we must assume that rearrangement has already taken place, that the elements $j \in \overline{J}$ are taken in the order $j_1 \dots j_m$, and that the matrix $\overline{B}_{\overline{J}}$ is used instead of the matrix $B_{\overline{J}}$. For example, $f'|_{\overline{J}}$ is now the row vector with components $\partial f / \partial x^j$, $j = j_1, \dots, j_m$:

$$f'|_{\overline{J}} = \{\partial f / \partial x^{j_1}, \partial f / \partial x^{j_2}, \dots, \partial f / \partial x^{j_m}\}.$$

Summarizing the above, we conclude that if the initial point x_0 is known then the initial basis $\overline{J}^0$ and matrix $B_{\overline{J}^0}$ may be efficiently constructed using only the Gauss–Jordan elimination method (as described above) applied to the matrix $A|_{\overline{J}(x_0)}$.

Let us now assume that, at some stage, there is a current point x, and the original matrix A has been transformed r times so that we have a current matrix $A^{(r)}$, a basis $\overline{J}$, a mapping $j(i)$ $(j = 1, \dots, m)$ and a matrix $\overline{B}_{\overline{J}}$ such that

$$\overline{B}_{\overline{J}} = T^{(r)} \dots T^{(1)}, \quad A^{(r)} = \overline{B}_{\overline{J}} A, \quad A^{(r)}_{j(i)} = e_i, \quad i = 1, \dots, m. \tag{1.65}$$

Suppose that the basis component x^{j_0} is transformed to zero. By virtue of the nondegeneracy condition, we can find not less than m linearly independent vectors among the A_l, $l \in \overline{J}(x) = \{j | x^j > 0\}$. Thus, if $i_0 = i(j_0)$, where $i(j)$ is the inverse mapping to $j(i)$ (in fact, each i is in one-to-one correspondence with an index j), then at least one of the elements $a^{(r)}_{i_0 l}$, $l \in \overline{J}(x) \backslash \overline{J}$ must be nonzero. In fact, by assumption, the number of elements of $\overline{J}(x)$ is not less than m. The index $j_0 \in \overline{J}$ for which $a_{i_0 j_0} = 1$ does not belong to this set. Furthermore, $a^{(r)}_{i_0 j} = 0$, $j \neq j_0$, $j \in \overline{J}$ by the construction of the matrix $A^{(r)}$ and the basis.

Thus, if $a_{i_0 l} = 0$ for all $l \in \overline{J}(x) \backslash \overline{J}$, then all the vectors $A^{(r)}_j$, $j \in \overline{J}(x)$ have zero i_0th component, i.e. they are actually of dimension $(m-1)$. At the same time, there are at least m of these and thus they are linearly dependent. This contradicts the nondegeneracy of the problem.

Therefore, if $i_0 = i(j_0)$, we can find an index $l \in \overline{J}(x) \backslash \overline{J}$, i.e. $l \in J_{\mathrm{a}}$, such that $a^{(r)}_{i_0 l} \neq 0$. Since $a^{(r)}_{i_0 j} = 0$, $j \neq j_0$, $j \in \overline{J}$, then $A^{(r)}_l$ is linearly independent of the A_j, $j \in \overline{J} \backslash \{j_0\}$; whence,

$$\overline{J} = (\overline{J} \backslash \{j_0\}) \cup \{l\}$$

is a set of indices for a new basis.

Beginning with the element $a^{(r)}_{il}$ and the matrix $A^{(r)}$ we now apply the transformation $T^{(r+1)} = T_{il}$ and obtain

$$A^{(r+1)} = T^{(r+1)} A^{(r)}, \quad \overline{B}^{(r+1)} = T^{(r+1)} \overline{B}^{(r)}.$$

As far as the mapping $j(i)$ is concerned, for $i \neq i_0$, it remains as before, while $j(i_0) = l$. Thus, we have shown that a change of basis may be implemented by means of Gauss–Jordan elimination.

To conclude, we consider the question of storing the data. There are two possible methods:

1. To store $A^{(r)}$, $\overline{B}^{(r)}$ and the mapping $j(i)$ (the latter clearly requires $2m$ memory locations and is not a problem). In this storage method, the matrix $A^{(r)}$ is transformed directly, i.e. if the element $a^{(r)}_{i_0 l}$ is chosen for the transformation then the rows $a^{(r+1)}_k$ of the matrix $A^{(r+1)}$ are obtained from the formulae:

$$\begin{aligned} a^{(r+1)}_k &= a^{(r)}_k + t^{(r)}_k a^{(r)}_{i_0}, \quad t^{(r)}_k = -a^{(r)}_{kl} / a^{(r)}_{i_0 l}, \quad k \neq i_0 \\ a^{(r+1)}_{i_0} &= t^{(r)}_{i_0} a^{(r)}_{i_0}, \quad t^{(r)}_{i_0} = 1 / a^{(r)}_{i_0 l}. \end{aligned}$$

   In addition, $\overline{B}^{(r+1)} = T^{(r+1)} \overline{B}^{(r)}$.

   This method of storage is simple and convenient for transformations but has the following disadvantages:

   a) Computational errors may accumulate.

b) This involves calculation of the nonzero elements of $A^{(r)}$; while the original matrix A may be comparatively sparse, $A^{(r)}$ could be heavily occupied for large r.

2. To store the original matrix A, the matrices $T^{(1)}, \dots, T^{(r)}$ and the function $j(i)$. Clearly, for the matrices $T^{(r)}$, only the nonzero elements $t_k^{(r)}$, $k = 1, \dots, m$ and the index i_r of the row upon which the transformation is based need be stored.

   The advantage of this storage technique is that, as a rule, the $t_k^{(r)}$ also include many zero elements, which need not be stored; thus memory is saved. Moreover, from time to time, knowing the current basis from the function $j(i)$ and the original matrix A, it is possible to delete all previously stored matrices $T^{(1)}, \dots, T^{(r)}$ and recalculate new matrices

$$T^{(i)} = T_{ij(i)}^{\cdot}, \quad i = 1, \dots, m$$

   where, naturally, each successive matrix $T^{(i+1)}$ is calculated from the elements of the previous matrix

$$A^{(i)} = T^{(i)} \dots T^{(1)} A.$$

   As we saw earlier,

$$B^{(r)} = \overline{B}_{\overline{J}} = [A_{j_1} \dots A_{j_m}]^{-1}, \quad A^{(r)} = B^{(r)} A;$$

   whence $A^{(r)}$ and $B^{(r)}$ do not depend on the sequence of matrices $T^{(1)}, \dots, T^{(r)}$, but only on the current basis $\overline{J}$ and the mapping $j(i)$.

   Recalculation of the matrices $T^{(r)}$ may be necessary because of the excessive accumulation of these or because of the need to prevent the accumulation of computational errors.

1.3.8 Algorithms for Simple Constraints. Generalization

The algorithm, described in Section 1.3.6 may be considerably simplified if there are only constraints on the signs of the variables, i.e. if we consider the problem

$$\min\{f(x) = 1/2(x, Cx) + (d, x) | x \geq 0\}.$$

In this case, there are no bases and the variables are simply divided into active and passive sets.

The algorithm takes the following form. Suppose that x_0 is the initial point.

0. If $\partial f / \partial x^j = 0$, $x_0^j > 0$ and $\partial f / \partial x^j \geq 0$, $x_0^j = 0$, then x_0 is a solution of the problem.

1. We set

$$J_{\mathrm{a}}^0 = \{j | x_0^j > 0 \text{ or } x_0^j = 0, \ \partial f / \partial x^j < 0\}, \quad J_{\mathrm{p}}^0 = \{1, \dots, n\} \backslash J_{\mathrm{a}}^0.$$

2. If $J_{\text{a}}^0 \neq \emptyset$, then for $k = 0, 1, \ldots$ we minimize the function $f(x)$ with respect to the variables x^j, $j \in J_{\text{a}}^k$ by the conjugate gradient method, verifying the conditions on the signs of the variables, as in the main algorithm. If during this process some coordinate x^j, $j \in J_{\text{a}}^k$ is transformed to zero, then

$$J_{\text{a}}^{k+1} = J_{\text{a}}^k \backslash \{j\}, \quad J_{\text{p}}^{k+1} = J_{\text{p}}^k \cup \{j\}$$

and we return to stage 2.

If for some J_{a}^k, the conjugate gradient method terminates, we have found a minimum with respect to the variables x^j, $j \in J_{\text{a}}^k$, the point obtained is taken as the initial point and we return to stage 1 of the algorithm.

The convergence of the algorithm follows from the proof that the general algorithm converges. However, in this case, the proof is considerably simpler and is based on the fact that during the process the set J_{p}^k is extended continually, whence stage 2 of the algorithm must terminate after a finite number of operations to find the minimum with respect to x^j, $j \in J_{\text{a}}^k$. Since the function $f(x)$ decreases throughout the process, the finite sets J_{a}^k at each stage of the process cannot be repeated. Since there is clearly a finite number of such sets the whole algorithm converges after a finite number of operations.

Here are a number of concluding remarks.

1. If any one of the variables x^j does not have a constraint on its sign, it is advisable to include it immediately in a basis, after which it is possible to ignore its sign during the process.

2. If the variables are bounded above, $0 \leq x^j \leq w^j$, then the algorithm may also be modified. In this case, both variables transformed to zero and those attaining the upper bound will be passive.

3. Instead of the above version of the conjugate gradient algorithm it is possible to use any other variant of it described in the literature. The only important requirement which must be satisfied is that the initial step of the process should be chosen in the direction of the anti-gradient.

2. The Linearization Method

2.1 The General Algorithm

In this section we consider a method for solving a general mathematical programming problem with no convexity assumptions on the functions involved. One important feature of this method is that it is able to take account of nonlinear equality-type constraints, which form a stumbling block for most other methods.

Suppose that it is required to minimize the function $f_0(x)$, $x \in E^n$, subject to the constraints

$$f_i(x) \leq 0, \quad i \in I^-, \qquad f_i(x) = 0, \quad i \in I^0, \tag{2.1}$$

where I^- and I^0 are finite sets of indices. We assume that all the functions $f_i(x)$ are continuously differentiable. The constraints for which the problem is studied will be discussed more fully below. At the point x_0 we replace all the constraints (2.1) and $f_0(x)$ by linear constraints, linearizing the $f_i(x)$ at the point x_0. Thus, we obtain a linear programming problem. It would be natural to take the solution of the linearized problem as the next approximation, as in Newton's method for solving systems of nonlinear equations. Unfortunately, this direct approach does not lead to the goal, since, more often than not, the auxiliary linear programming problem does not have a solution. Thus, we have to impose constraints on the increment of the vector x at the point x_0 so that the solution of the linearized problem at the point x_0 is not too far away from x_0 and remains in the neighbourhood of x_0 in which the linearization is valid. We shall do this below by adding a quadratic term to the linearized goal function.

We note that each equation $f_i(x) = 0$ is equivalent to two inequalities, $f_i(x) \leq 0$ and $-f_i(x) \leq 0$. Thus, we may restrict our attention to the case in which all the constraints are in the form of inequalities. This restriction is convenient, at least as far as theoretical justification of the algorithm is concerned, although the doubling of the number of inequalities may be computationally inconvenient. In what follows, we shall study the theoretical justification of the algorithm for the problem of minimizing $f_0(x)$ subject to the constraints

$$f_i(x) \leq 0, \quad i \in I. \tag{2.2}$$

Modification of the algorithm for the general problem (2.1) will be discussed separately.

Thus, without loss of generality, we shall study the algorithm for problem (2.2). Clearly, we may always assume that the inequalities of (2.2) include the trivial inequality $0 \leq 0$. Thus, we shall assume that one of the functions $f_i(x)$, $i \in I$ is identically zero.

2.1.1 Main Assumptions

We set

$$\begin{aligned} F(x) &= \max_{i \in I} f_i(x) \\ I_\delta(x) &= \{i \in I | f_i(x) \geq F(x) - \delta\}, \quad \delta \geq 0. \end{aligned} \tag{2.3}$$

Because of the previous assumption, $F(x) \geq 0$ for all x. We assume that there exist constants $N > 0$, $\delta > 0$, such that

a) The set

$$\Omega_N = \{x | f_0(x) + NF(x) \leq C_0\}, \quad C_0 = f_0(x_0) + NF(x_0)$$

is bounded.

b) The gradients of the functions $f_i(x)$, $i \in \{0\} \cup I$, on Ω_N satisfy a Lipschitz condition, i.e.

$$\|f_i'(x_1) - f_i'(x_2)\| \leq L\|x_1 - x_2\|;$$

c) The quadratic programming problem

$$\begin{aligned} &\min((f_0'(x), p) + 1/2\|p\|^2) \\ &(f_i'(x), p) + f_i(x) \leq 0, \quad i \in I_\delta(x) \end{aligned} \tag{2.4}$$

is solvable over $p \in E^n$ for any $x \in \Omega_N$ and there exist Lagrange multipliers $u^i(x)$, $i \in I_\delta(x)$, such that $\sum_{i \in I_\delta(x)} u^i(x) \leq N$. Here and throughout this section, $\|p\|$ denotes the Euclidean norm of the vector p.

In what follows, we shall denote the solution of problem (2.4) by $p(x)$ and the Lagrange multipliers by $u^i(x)$, $i \in I_\delta(x)$.

2.1.2 Formulation of the Algorithm

Suppose that x_0 is an initial approximation and that ϵ is chosen such that $0 < \epsilon < 1$. Suppose that the algorithm has already been applied to obtain the point x_k. The next approximation is constructed in two stages.

1. We solve problem (2.4) for $x = x_k$, where its solution is the vector $p_k = p(x_k)$.

2. We find the first value of $i = 0, 1, \ldots$ for which the inequality

$$f_0(x_k+(1/2)^i p_k) + NF(x_k+(1/2)^i p_k) \le f_0(x_k) + NF(x_k) - (1/2)^i \epsilon \|p_k\|^2$$

is satisfied. If this inequality is first satisfied for $i = i_0$, we set

$$\alpha_k = 2^{-i_0}, \quad x_{k+1} = x_k + \alpha_k p_k.$$

Thus, at each step, the following inequality is satisfied:

$$f(x_{k+1}) + NF(x_{k+1}) \le f(x_k) + NF(x_k) - \alpha_k \epsilon \|p_k\|^2. \tag{2.5}$$

2.1.3 Convergence of the Algorithm

We shall now show that the choice of the step α_k at each iteration is completed after dividing by two finitely many times and prove that the algorithm converges.

From the results of Section 1.2, it follows that $p(x)$ is a solution of problem (2.4) if and only if there exist $u^i(x) \ge 0$, $i \in I_\delta(x)$, such that

$$\begin{aligned} f_0'(x) + (p(x))^* + \sum_{i \in I_\delta(x)} u^i(x) f_i'(x) &= 0 \\ u^i(x)((f_i'(x), p(x)) + f_i(x)) &= 0, \quad i \in I_\delta(x). \end{aligned} \tag{2.6}$$

We recall that, in our notation, the gradient $f'(x)$ is a row vector. Since $p(x)$ is a column vector, it occurs in (2.6) in transposed form, where it appears as the gradient of the function $\frac{1}{2}\|p\|^2$. Thus,

$$\begin{aligned} (f_0'(x), p(x)) &= -\sum_{i \in I_\delta(x)} u^i(x)(f_i'(x), p(x)) - \|p(x)\|^2 \\ &= \sum_{i \in I_\delta(x)} u^i(x) f_i(x) - \|p(x)\|^2 \end{aligned} \tag{2.7}$$

Lemma 2.1 *The point x satisfies the inequalities (2.2) and the necessary conditions for a minimum of $f_0(x)$ subject to the constraints (2.2) are satisfied there if and only if the equation $p(x) = 0$ is satisfied.*

Proof. Suppose that the point x satisfies (2.2) and that the necessary conditions for a minimum of $f_0(x)$ are satisfied there. Then, there exist numbers $u^i \ge 0$, $i \in I$, such that

$$f_0'(x) + \sum_{i \in I} u^i f_i'(x) = 0, \quad u^i f_i(x) = 0, \quad i \in I. \tag{2.8}$$

If x satisfies (2.2) then $F(x) = 0$ and thus $I_0(x)$ is the set of i for which $f_i(x) = 0$. In addition, by virtue of the second equation of (2.8), $u^i = 0$ if $f_i(x) < 0$, i.e. if $i \notin I_0(x)$. Thus, since $I_\delta(x) \supset I_0(x)$, (2.8) may be rewritten in the form

$$f_0'(x) + \sum_{i \in I_\delta(x)} u^i f_i'(x) = 0, \quad u^i f_i(x) = 0, \quad i \in I_\delta(x).$$

But, comparison of the above equations with (2.6) shows that the vector $p = 0$ is a solution of problem (2.4), since, for $p = 0$ all the constraints of (2.4) are satisfied (since those of (2.2) are satisfied), while the satisfaction of the equations of (2.6) for $p = 0$ is a necessary and sufficient condition for the vector $p = 0$ to be a solution of (2.4).

Suppose now that $p(x) = 0$. Then the constraints of problem (2.4) are satisfied for $p = 0$, i.e. $f_i(x) \leq 0$, $i \in I_\delta(x)$. Since for $i \notin I_\delta(x)$

$$f_i(x) \leq F(x) - \delta \leq f_j(x) \leq 0$$

where $j \in I_\delta(x)$, the point x satisfies all the constraints of (2.2). In addition, for $p = 0$, equation (2.6) becomes (2.8) if we assume that $u^i = 0$, $i \notin I_\delta(x)$. Thus, the necessary conditions for a minimum of $f_0(x)$ subject to the constraints (2.2) are also satisfied. This completes the proof □

We now estimate the change in all the functions involved, given a displacement of x_k in the direction p_k. For $i \in I_\delta(x_k)$, using Taylor's formula, we obtain

$$f_i(x_k + \alpha p_k) = f_i(x_k) + \alpha(p_k, f_i'(x_k)) + \alpha(p_k, f_i'(\theta_i) - f_i'(x_k))$$

where $\theta_i = x_k + \alpha \xi_i p_k$, $0 \leq \xi_i \leq 1$. Since p_k is a solution of (2.4) for $x = x_k$, we have

$$\begin{aligned} f_i(x_k + \alpha p_k) &\leq f_i(x_k) - \alpha f_i(x_k) + \alpha^2 \|p_k\|^2 L \\ &\leq (1 - \alpha) f_i(x_k) + \alpha^2 \|p_k\|^2 L \end{aligned} \tag{2.9}$$

(in deriving (2.9), we have used the fact that the gradients of $f_i(x)$ satisfy a Lipschitz condition).

For $i \notin I_\delta(x)$

$$\begin{aligned} f_i(x_k + \alpha p_k) &= f_i(x_k) + \alpha(p_k, f_i'(\theta_i)) \\ &\leq F(x_k) - \delta + \alpha K \|p_k\| \end{aligned} \tag{2.10}$$

where K is the value of the bound on $\|f_i'(x)\|$ on Ω_N.

Since

$$(1 - \alpha) F(x_k) \geq F(x_k) - \delta + \alpha K \|p_k\|$$

for

$$\alpha \leq 1, \quad 0 \leq \alpha \leq \frac{\delta}{F(x_k) + K\|p_k\|} \tag{2.11}$$

it follows from (2.9) and (2.10) that the following inequality

$$f_i(x_k + \alpha p_k) \leq (1 - \alpha) F(x_k) + \alpha^2 L \|p_k\|^2 \tag{2.12}$$

holds for all i.

Analogously to the above estimation we obtain

$$\begin{aligned} f_0(x_k+\alpha p_k) &= f_0(x_k)+\alpha(p_k, f_0'(x_k))+\alpha(p_k, f_0'(\theta_0)-f_0(x_k)) \\ \theta_0 &= x_k+\alpha\xi_0 p_k, \quad 0\le\xi_0\le 1. \end{aligned}$$

Using (2.7) and the Lipschitz condition for the gradients, we obtain

$$\begin{aligned} f_0(x_k+\alpha p_k) \le\; & f_0(x_k)+\alpha\Big(\sum_{i\in I_\delta(x_k)} u^i(x_k)f_i(x_k)\Big) \\ & -\alpha\|p_k\|^2+\alpha^2 L\|p_k\|^2. \end{aligned}$$

From the above, together with (2.12), it follows that

$$\begin{aligned} f_0(x_k+\alpha p_k)+NF(x_k+\alpha p_k) \le\; & f_0(x_k)+NF(x_k) \\ & +\alpha\Big(\sum_{i\in I_\delta(x_k)} u^i(x_k)f_i(x_k)-NF(x_k)\Big) \\ & -\alpha\|p_k\|^2+\alpha^2(N+1)L\|p_k\|^2. \end{aligned} \tag{2.13}$$

We now recall that $u^i(x_k)\ge 0$, $F(x_k)\ge 0$ and

$$\sum_{i\in I_\delta(x_k)} u^i(x_k)\le N.$$

Thus,

$$\sum_{i\in I_\delta(x_k)} u^i(x_k)f_i(x_k)-NF(x_k)\le 0.$$

But then (2.13) may be rewritten in the form

$$\begin{aligned} f_0(x_k+\alpha p_k)+NF(x_k+\alpha p_k) \le\; & f_0(x)+NF(x_k) \\ & -\alpha\|p_k\|^2(1-\alpha(N+1)L). \end{aligned}$$

If

$$0\le\alpha\le\frac{1-\epsilon}{(N+1)L} \tag{2.14}$$

then

$$f_0(x_k+\alpha p_k)+NF(x_k+\alpha p_k)\le f_0(x)+NF(x_k)-\alpha\epsilon\|p_k\|^2. \tag{2.15}$$

Thus, if

$$\begin{aligned} 0 &\le \alpha\le\overline{\alpha}_k, \\ \overline{\alpha}_k &= \min\left(1, \frac{\delta}{F(x_k)+K\|p_k\|}, \frac{1-\epsilon}{(N+1)L}\right) \end{aligned}$$

then inequality (2.15) is satisfied. But this means that inequality (2.5) will be satisfied after a finite number of trials $\alpha=2^{-i}$, $i=0,1,\ldots$, and the inequality

$$\alpha_k > \overline{\alpha_k}/2 \tag{2.16}$$

will be satisfied.

We next prove the following theorem about the convergence of the process.

Theorem 2.1 *If the assumptions of Section 2.1.1 are satisfied then the process has the following properties:*

a) $F(x_k) \to 0$ *as* $k \to \infty$*;*

b) the inequalities of (2.2) are satisfied at any limit point x_* *of the sequence* x_k, $k = 0, 1, \ldots$ *as are the necessary conditions for a minimum of* $f_0(x)$ *subject to the constraints (2.2).*

Remark. The convergence of $F(x_k)$ to zero again means that the sequence x_k satisfies the constraints (2.2).

Proof. All the points x_k lie in the region Ω_N since, by virtue of (2.15), the function $f_0(x) + NF(x)$ decreases from step to step. In addition, since Ω_N is compact, $f_0(x) + NF(x)$ is bounded on this set, because it is a continuous function. Whence it follows that

$$\alpha_k \|p_k\|^2 \to 0 \text{ as } k \to \infty \tag{2.17}$$

since otherwise $f_0(x) + NF(x)$ would decrease in an unbounded manner for the sequence x_k.

We shall show that $p_k \to 0$. Clearly, if $p_k \not\to 0$, then it follows from (2.17) that $\alpha_k \to 0$ for some subsequence of indices k. But in this case it follows from (2.16) and the expression for $\overline{\alpha}_k$ that for large k

$$\alpha_k \geq \frac{1}{2}\overline{\alpha}_k = \frac{\delta}{2(F(x_k) + K\|p_k\|)};$$

whence the right-hand side of the previous inequality must tend to zero. Since $F(x)$ is a continuous function on the compact set Ω_N, $F(x)$ is bounded above and the expression $\delta/(F(x_k) + K\|p_k\|)$ may converge to zero only if $\|p_k\| \to +\infty$. But, from (2.6),

$$\|p_k\| = \|f_0'(x_k) + \sum_{i \in I_\delta(x_k)} u^i(x_k) f_i'(x_k)\| \leq K(N+1).$$

Thus, we have obtained a contradiction to the assumption that $p_k \not\to 0$.

By definition of p_k, the following equations are satisfied:

$$(f_i'(x_k), p_k) + f_i(x_k) \leq 0, \quad i \in I_\delta(x_k).$$

Thus,

$$f_i(x_k) \leq -(f_i'(x_k), p_k) \leq K\|p_k\|, \quad i \in I_\delta(x_k).$$

But $f_j(x_k) \leq f_i(x_k)$, $j \notin I_\delta(x_k)$, $i \in I_\delta(x_k)$. Whence

$$F(x_k) = \max_{i \in I} f_i(x_k) \leq K\|p_k\|.$$

Thus, $F(x_k) \to 0$ as $k \to \infty$ since $F(x_k) \geq 0$. We now set $u^i(x) = 0$, $i \notin I_\delta(x)$. Then for the sequence x_k, equation (2.6) may be rewritten in the form

$$\begin{aligned} f_0'(x_k) + p_k^* + \sum_{i \in I} u^i(x_k) f_i'(x_k) &= 0 \\ u^i(x_k)((f_i'(x_k), p_k) + f_i(x_k)) &= 0, \quad i \in I \end{aligned} \tag{2.18}$$

Suppose now that x_* is a limit point of the sequence $\{x_k\}$. Since $x_k \in \Omega_N$, which is compact, limit points always exist. Without loss of generality, we may assume that $x_k \to x_*$. Moreover, since $u^i(x) \geq 0, i \in I$ and their sum is bounded, we may assume that $u^i(x_k) \to u^i$ as $k \to \infty$.

Passing to the limit in (2.18), we obtain

$$f_0'(x_*) + \sum_{i \in I} u^i f_i'(x_*) = 0, \quad u^i f_i(x_*) = 0, \quad i \in I.$$

In addition, $u^i \geq 0$, since $u^i(x_k) \geq 0$ and the point x_* satisfies all the constraints of (2.2). Clearly, $f_i(x_k) \leq F(x_k)$ and $F(x_k) \to 0$; whence, taking limits, we obtain that $f_i(x_*) \leq 0$. Thus, we conclude that the necessary conditions for a minimum are satisfied at the point x_*. The proof of the theorem is complete. □

Corollary *If the unique point at which the necessary conditions for a minimum are satisfied is a minimum then the sequence generated by the algorithm converges to the minimum of $f_0(x)$ subject to the constraints (2.2).*

In this case, it is clear from Theorem 2.1 that the unique limit point of the sequence x_k can only be a minimum.

2.1.4 Computational Aspects

The main operation, which involves considerable computation at each step of the algorithm, is the resolution of problem (2.4). This is a quadratic programming problem. When choosing a method for solving this problem, one should bear in mind that since problem (2.4) is an auxiliary problem, its solution must be obtained in a finite number of steps. In addition, since the previous constant N is generally unknown, corresponding Lagrange multipliers $u^i(x)$ may be obtained to ensure that N is chosen correctly when solving problem (2.4). Under these conditions, when solving problem (2.4), it is advisable to move to the dual problem, which may be solved using the conjugate gradient method described in Section 1.3.8.

We shall construct the dual problem for problem (2.4). According to Section 1.2, the goal function of the dual problem has the form

$$\varphi(u) = \min_p[(f_0'(x), p) + (1/2)\|p\|^2 + \sum_{i \in I_\delta(x)} u^i((f_i'(x), p) + f_i(x))]. \qquad (2.19)$$

Equating the derivative with respect to p of the right-hand side of the above equation to zero, we find that a minimum is attained for

$$p = -f_0'(x) - \sum_{i \in I_\delta(x)} u^i f_i'(x). \qquad (2.20)$$

Thus, the point p is uniquely determined by the vector u with components u^i, $i \in I_\delta(x)$.

Substituting (2.20) into the right-hand side of (2.19), we obtain

$$\varphi(u) = -(1/2)\|f_0'(x) + \sum_{i \in I_\delta(x)} u^i f_i'(x)\|^2 + \sum_{i \in I_\delta(x)} u^i f_i(x). \qquad (2.21)$$

Thus, we have calculated the goal function of the dual problem. The dual problem itself then involves maximizing $\varphi(u)$ subject to the constraints $u^i \geq 0$, $i \in I_\delta(x)$.

Thus, we have obtained a problem involving maximization of a quadratic form subject to simple constraints, which may easily be solved using the conjugate gradient method (Section 1.3.8). As a result of solving this, we obtain Lagrange multipliers $u^i(x)$ (solutions of the dual problem), which, according to Section 1.3 when substituted in (2.20) give the vector $p(x)$, the solution of the original problem.

Another problem is that of choosing the constants N and δ. Generally speaking, the value of N is unknown. It is disadvantageous to choose it too large, since, because of formula (2.14), this may lead to a significant reduction of the step. Thus, it is best to estimate it as part of the algorithm. For example, if at some step we have

$$N \leq \sum_{i \in I_\delta(x_k)} u^i(x_k) \qquad (2.22)$$

then N should be changed and replaced by

$$N = 2 \sum_{i \in I_\delta(x_k)} u^i(x_k). \qquad (2.23)$$

Practical experience shows that correction in this way leads to success. Moreover, based on theoretical considerations, it is clear that if x_k is sufficiently close to a limit point then in the regular case the $u^i(x_k)$ will be close to the Lagrange multipliers at the point x_* which is a solution of the problem, and thus formula (2.23) leads to success. Details of the behaviour of the Lagrange multipliers $u^i(x_k)$ will be discussed below.

As far as the value of δ is concerned, it should be decreased if at some stage problem (2.4) appears unsolvable. However, experience of solving such problems shows that, in practice, δ should be taken to be as large as possible, taking into account all the constraints of the original problem when solving the auxiliary problem, subject to the machine being able to store this value.

2.1.5 Generalizations

At the beginning of this section, we mentioned the fact that in the case of equality-type constraints (i.e. constraints of the form (2.1)) the problem reduces to the form (2.2) in which each equation is replaced by two inequalities. Thus, we apply the algorithm to the general problem (2.1). Here, we need only take into account the fact that, if for some x

$$f_i(x) \geq F(x) - \delta, \quad -f_i(x) \geq F(x) - \delta, \tag{2.24}$$

where $i \in I_0$, then the system (2.4) involves two inequalities

$$(f_i'(x), p) + f_i(x) \leq 0, \quad -(f_i'(x), p) - f_i(x) \leq 0, \tag{2.25}$$

which are equivalent to the single equation

$$(f_i'(x), p) + f_i(x) = 0. \tag{2.26}$$

Thus, it is advisable to take this fact into account when solving the auxiliary problem and to replace pairs of inequalities of the form (2.25) in (2.4) by a single equation (2.26). When passing to the dual problem, this implies that the corresponding multiplier u^i has an arbitrary sign; however, this does not prevent the possible application of the conjugate gradient method (Section 1.3.8).

Let us now assume that in addition to the constraints of (2.2), the original problem is also subject to the condition that the minimum x belongs to some set X with a simple structure. In this case, it is advisable that the approximations obtained should lie in the set X. We shall describe how the algorithm is modified in this case. As before, without loss of generality, we consider only cases when the constraints are all inequalities.

Thus, suppose that it is required to minimize $f_0(x)$, $x \in E^n$ subject to the constraints

$$f_i(x), \quad i \in I, \quad x \in X \tag{2.27}$$

where I is a finite set of indices and X is a convex closed set. We suppose that there exists an index i such that $f_i(x) = 0$.

We also assume that there exist constants $N > 0$ and $\delta > 0$ such that the following conditions are satisfied:

a) the sets

$$\begin{aligned} \Omega_N &= \{x | f_0(x) + NF(x) \leq C_0, \ x \in X\} \\ C_0 &= f_0(x_0) + NF(x_0) \end{aligned}$$

are bounded and the initial approximation x_0 belongs to X;

b) the gradients of the functions $f_i(x)$, $i \in \{0\} \cup I$ on Ω_N satisfy a Lipschitz condition, i.e.

$$\|f_i'(x_1) - f_i'(x_2)\| \leq L\|x_1 - x_2\|;$$

c) the problem

$$\begin{aligned}&\min(f_0'(x),p)+1/2\|p\|^2\\&(f_i'(x),p)+f_i(x)\le 0,\quad i\in I_\delta(x),\quad x+p\in X\end{aligned}\tag{2.28}$$

is solvable over all p for any $x\in\Omega_N$ and there exist Lagrange multipliers $u^i(x)$, $i\in I_\delta(x)$, such that

$$\sum_{i\in I_\delta(x)} u^i(x)\le N.$$

Remark. We recall that the Lagrange multipliers for problem (2.28) are nonnegative numbers which satisfy the condition

$$\begin{aligned}&(f_0'(x),p(x))+(p(x),p(x))\\&\quad+\sum_{i\in I_\delta(x)} u^i(x)[(f_i'(x),p(x))+f_i(x)]\le (f_0'(x),p)+(p(x),p)\\&\quad+\sum_{i\in I_\delta(x)} u^i(x)[(f_i'(x),p)+f_i(x)]\end{aligned}\tag{2.29}$$

for all p satisfying the condition

$$x+p\in X.\tag{2.30}$$

In addition,

$$u^i(x)[(f_i'(x),p(x))+f_i(x)]=0,\quad i\in I_\delta(x).\tag{2.31}$$

Thus, condition c) assumes not only that the auxiliary problem (2.28) is solvable, but also that the necessary and sufficient conditions of the Kuhn–Tucker Theorem are satisfied at its minimum $p=p(x)$.

The algorithm to solve problem (2.27) is then constructed in exactly the same way as described in Section 2.1.2, except that now the p_k are taken to be the vectors $p(x_k)$, the solutions of the new auxiliary problem (2.28).

We shall prove that the algorithm converges, i.e. that the conclusions of Theorem 2.1 are satisfied and that $x_k\in X$ for all k. From the last assertion, it follows, in particular, that every limit point of the sequence x_k lies in X. Since the proof of convergence only differs from the proof of Theorem 2.1 in a few places, we do not need to repeat the proof in detail. We note only the main differences.

Firstly, since $x_k+p_k\in X$ and X is convex, then $x_k+\alpha p_k\in X$ for all α between 0 and 1. Thus, if $x_k\in X$ then $x_{k+1}\in X$. Since $x_0\in X$, by assumption, the whole sequence $\{x_k\}_{k=0}^{\infty}$ lies in X.

Secondly, from (2.29)–(2.31), for $p=0$, it follows that

$$(f_0'(x),p(x))+\|p(x)\|^2\le\sum_{i\in I_\delta(x)} u^i(x)f_i(x),$$

in other words

$$(f_0'(x), p(x)) \le \sum_{i \in I_\delta(x)} u^i f_i(x) - \|p(x)\|^2. \qquad (2.32)$$

This inequality replaces equation (2.7) which was used to obtain the bound (2.13). All the remaining calculations to obtain the bound remain unchanged.

Finally, if $p(x_*) = 0$, then from (2.29)–(2.31), it follows that the conditions

$$\begin{aligned} (f_0'(x_*), p) + \sum_{i \in I_\delta(x)} u^i(x_*)(f_i'(x_*), p) &\ge 0 \\ x_* + p \in X, \ \ u^i(x_*) f_i(x_*) &= 0, \ \ i \in I_\delta(x*) \end{aligned} \qquad (2.33)$$

are satisfied. Moreover, in this case, it follows from (2.28) that

$$f_i(x_*) \le 0, \ \ i \in I_\delta(x_*), \ \ x_* \in X$$

and in addition, clearly

$$f_i(x_*) < 0, \ \ i \notin I_\delta(x_*).$$

Thus, the point x_* satisfies all the constraints of (2.27) and conditions (2.33) show that the necessary conditions for an extremum are satisfied at this point.

Therefore, as before, we have shown that if $p(x_*) = 0$ then the necessary conditions for an extremum are satisfied at the point x_*. It is easy to show the converse, namely that the condition $p(x) = 0$ is a necessary and sufficient condition for the point x to be susceptible to being an extremum.

The proof that each limit point x_* of the sequence x_k $k = 0, 1, \ldots,$ satisfies the necessary conditions for an extremum is the same as that of Theorem 2.1, where in passing to the limit, the conditions (2.29)–(2.31) satisfied at x_k become equation (2.33) which is satisfied at the limit point.

2.1.6 The Linear Programming Problem

Suppose now that in problem (2.2), all the functions $f_0(x)$, $f_i(x)$, $i \in I$ are linear. Thus, we obtain a linear programming problem. While the above algorithm is more important for the nonlinear case, it is also reasonable to apply it to linear programming problems. In particular, if the set I contains a large number of indices, then we have a linear programming problem with a large number of constraints. At the same time, for small δ, the auxiliary problem (2.4) will only have a small number of constraints, so that the general algorithm reduces to solving a series of simpler problems. In addition, unlike in the simplex method, computational errors are not accumulated in this method, since the original matrix of constraints is not transformed from step to step.

For linear programming problems, conditions a) and c) (condition b) is automatically satisfied) of the main assumptions are unnecessarily strict for convergence of the algorithm. We shall not discuss the conditions for the convergence for linear programming problems here, since our main aim is to derive an algorithm for the nonlinear case. In what follows we show that, at least when the previous assumptions a) and c) are satisfied in the linear programming problem, the algorithm will converge after a finite number of steps. This fact will in some way characterize the rate of convergence of our algorithm.

Theorem 2.2 *Suppose that assumptions a) and c) of Section 2.1.1 are satisfied and that all the functions $f_0(x)$, $f_i(x)$ of problem (2.2) are of the form $f_i(x) = (a_i, x) - b_i$, then the algorithm of Section 2.1.2 converges after a finite number of steps.*

Proof. We note immediately that in this case the step α_k is equal to 1 for k sufficiently large. Clearly, since all the $f_i(x)$ are linear, the Lipschitz constant L may simply be taken to equal zero. Thus, from the formula of Section 2.1.3 for $\overline{\alpha}_k$, it follows that

$$\begin{aligned} \overline{\alpha}_k &= \min\left(1, \frac{\delta}{F(x_k) + K\|p_k\|}, \frac{1-\epsilon}{(N+1)L}\right) \\ &= \min\left(1, \frac{\delta}{F(x_k) + K\|p_k\|}\right). \end{aligned} \tag{2.34}$$

But we showed earlier that $F(x_k) \to 0$ as $\|p_k\| \to 0$. Thus, for k sufficiently large, we have $\delta/(F(x_k) + K\|p_k\|) \geq 1$ and $\overline{\alpha}_k = 1$. But $\overline{\alpha}_k$ was constructed in such a way that inequality (2.15) is satisfied for $\overline{\alpha} = \overline{\alpha}_k$. Since in each iteration the selection of α_k begins by division of $\alpha = 1$ by two, it follows that inequality (2.5), which determines the choice of α, is immediately satisfied with no additional division and the step α_k is simply equal to 1.

Suppose now that x_* is an arbitrary limit point of the sequence x_k obtained using the algorithm. As we know, this point is a solution of problem (2.2) since it satisfies all the constraints of the problem and in addition, according to Theorem 2.1, the necessary conditions for a minimum (which in the case of a linear programming problem are also sufficient conditions) are satisfied at this point.

We set

$$I_0(x_*) = \{i \in I | f_i(x_*) = 0\}. \tag{2.35}$$

Then $f_i(x_*) < 0$ for $i \notin I_0(x_*)$, so that

$$\epsilon_0 = \max_{i \notin I_0(x_*)} f_i(x_*) < 0. \tag{2.36}$$

To simplify the subsequent notation, without loss of generality, we shall assume that the whole sequence x_k converges to x_*.

Let us consider the auxiliary problem (2.4) at the points of the sequence x_k:

$$\begin{aligned} &\min(f_0'(x_k), p) + 1/2\|p\|^2 \\ &(f_i'(x_k), p) + f_i(x_k) \leq 0, \quad i \in I_\delta(x) \end{aligned} \tag{2.37}$$

where $p_k = p(x_k)$ is a solution of (2.4). We denote the corresponding Lagrange multipliers by u_k^i, $i \in I_\delta(x_k)$, so that

$$u_k^i[(f_i'(x_k), p_k) + f_i(x_k)] = 0. \tag{2.38}$$

We now show that $I_0(x_*) \subset I_\delta(x_k)$ for all k sufficiently large. Clearly, if $i \notin I_\delta(x_k)$, then

$$f_i(x_k) < F(x_k) - \delta.$$

Passing to the limit as $k \to \infty$, taking into account the fact that $F(x_k) \to 0$, we obtain $f_i(x_*) \le -\delta$, which would contradict the fact that i belongs to the set $I_0(x_*)$.

We now denote

$$\widetilde{I}(x_k) = \{i \in I_\delta(x_k) | u_k^i > 0\}.$$

The next assertion is that for all sufficiently large indices k,

$$\widetilde{I}(x_k) \subseteq I_0(x_*). \tag{2.39}$$

Clearly, if $i \notin I_0(x_*)$, then $f_i(x_*) \le \epsilon_0$. Since $p_k \to 0$, the $f_i'(x_k)$ are bounded and $x_k \to x_*$, then for large k,

$$|(f_i'(x_k), p_k)| \le -\epsilon_0/4, \quad f_i(x_k) \le \epsilon_0/2,$$

whence

$$(f_i'(x_k), p_k) + f_i(x_k) \le \epsilon_0/4 < 0.$$

Thus, if $u_k^i > 0$, then

$$u_k^i[(f_i'(x_k), p_k) + f_i(x_k)] < 0,$$

which contradicts (2.38).

Remark. None of the arguments given use the linearity of the $f_i(x)$; thus, the assertions that $I_0(x_*) \subset I_\delta(x_k)$ and $\widetilde{I}(x_k) \subset I_0(x_*)$ are true in the general case of a nonlinear problem. They will be used in what follows.

As shown in Section 2.1.4, the dual problem of the auxiliary problem (2.37) involves maximizing the function of (2.21) subject to the constraints $u^i \ge 0$, $i \in I_\delta(x_k)$. Here, the Lagrange multipliers u_k^i are a solution of the dual problem and the optimal values of the direct and the dual problems are the same, i.e.

$$(f_0'(x_k), p_k) + 1/2\|p_k\|^2 = -1/2\|f_0'(x_k) + \sum_{i \in I_\delta(x_k)} u_k^i f_i'(x_k)\|^2 + \sum_{i \in I_\delta(x_k)} u_k^i f_i(x_k).$$

Since $p_k \to 0$, the left-hand side of the above equation tends to zero, whence

$$-1/2\|f_0'(x_k) + \sum_{i \in I_\delta(x_k)} u_k^i f_i'(x_k)\|^2 + \sum_{i \in I_\delta(x_k)} u_k^i f_i(x_k) \to 0. \tag{2.40}$$

We now note that $u_k^i > 0$ only if $i \in \widetilde{I}(x_k)$. In addition, $f_i(x) = (a_i, x) - b_i$, $i \in \{0\} \cup I$, so that $f_i'(x) = a_i$ and is independent of x. Thus, (2.40) may be rewritten in the form

$$-1/2\|a_0 + \sum_{i \in \widetilde{I}(x_k)} u_k^i a_i\|^2 + \sum_{i \in \widetilde{I}(x_k)} u_k^i f_i(x_k) \to 0.$$

But, as shown above, $\widetilde{I}(x_k) \subset I_0(x_*)$, whence $f_i(x_k) \to f_i(x_*) = 0$, because $f_i(x_*) = 0$ for $i \in I_0(x_*)$, by definition. Thus,

$$-1/2\|a_0 + \sum_{i \in \widetilde{I}(x_k)} u_k^i a_i\|^2 \to 0.$$

But

$$-1/2\|a_0 + \sum_{i \in \widetilde{I}(x_k)} u_k^i a_i\|^2 \leq \max_{u^i \geq 0, i \in \widetilde{I}(x_k)} -1/2\|a_0 + \sum_{i \in \widetilde{I}(x_k)} u^i a_i\|^2 \leq 0. \quad (2.41)$$

We set

$$\omega(\overline{I}) = \max_{u^i \geq 0, i \in \overline{I}} -\|a_0 + \sum_{i \in \overline{I}} u^i a_i\|^2$$

where $\omega(\overline{I})$ is a function defined on a set of indices $\overline{I}$, $\overline{I} \subset I$. Since $\overline{I} \subset I$, this function may take only a finite number of values. It follows from (2.41) that $\omega(\widetilde{I}(x_k)) \to 0$. But this means that

$$\omega(\widetilde{I}(x_k)) = 0 \quad (2.42)$$

for all k sufficiently large, because, as mentioned above, $\omega(\overline{I})$ takes only finitely many values. We now choose k sufficiently large that $\alpha_k = 1$, condition (2.42) is satisfied and $\widetilde{I}(x_k) \subset I_0(x_*)$. Since $\alpha_k = 1$, $x_{k+1} = x_k + p_k$. Since $x_k \to x_*$ and $p_k \to 0$, we may assume that

$$f_i(x_{k+1}) \leq \epsilon_0/2 < 0, \quad i \notin I_0(x_*). \quad (2.43)$$

We now return to consider the auxiliary problem (2.37). Since p_k satisfies the constraints (2.37) and the $f_i(x)$ are linear, we have

$$f_i(x_{k+1}) = (f_i'(x_k), p_k) + f_i(x_k) \leq 0 \quad (2.44)$$

where $i \in I_\delta(x_k)$, whence, for $i \in I_0(x_*)$, since $I_0(x_*) \subset I_\delta(x_k)$. Thus, we have proved that x_{k+1} satisfies all the constraints of problem (2.2).

We now show that x_{k+1} is actually a solution of problem (2.2). Clearly, from (2.38) and the definition of the set $\widetilde{I}(x_k)$, it follows that

$$f_i(x_{k+1}) = 0, \quad i \in \widetilde{I}(x_k). \quad (2.45)$$

But (2.42) implies that there exist numbers $u_0^i \geq 0$, $i \in \widetilde{I}(x_k)$, such that

$$a_0 + \sum_{i \in \widetilde{I}(x_k)} u_0^i a_i = 0. \tag{2.46}$$

Then, setting $u_0^i = 0$, $i \notin \widetilde{I}(x_k)$, we obtain that there exist numbers $u_0^i \geq 0$, such that the conditions

$$a_0 + \sum_{i \in I} u_0^i a_i = 0, \quad u_0^i f_i(x_{k+1}) = 0$$

are satisfied. But the above equations (see Section 1.2) are necessary and sufficient conditions for the point x_{k+1} to be a solution of the linear programming problem.

Thus, the algorithm clearly leads to a solution after a finite number of steps and the proof of the theorem is complete. □

2.1.7 The Linearization Method with Equality-Type Constraints

Let us now consider the problem

$$\min\{f_0(x) | f_i(x) = 0, \ i = 1, \ldots, m\}. \tag{2.47}$$

For this problem

$$\begin{aligned} F(x) &= \max\{0, f_1(x), \ldots, f_m(x), -f_1(x), \ldots, -f_m(x)\} \\ &= \max_{1 \leq i \leq m} |f_i(x)| \end{aligned}$$

We choose $\delta = +\infty$, so that the auxiliary quadratic programming problem includes all the constraints and is of the form

$$\min_p\{(f_0'(x), p) + 1/2\|p\|^2 | (f_i'(x), p) + f_i(x) = 0, \ i = 1, \ldots, m\}. \tag{2.48}$$

The results of Section 1.3.4 may be applied to this problem. From formulae (1.48) and (1.49), with in the present case

$$C = I_n, \quad d = (f_0'(x))^*, \quad b = -f(x), \quad A = f'(x)$$

where $f(x) \in \mathbb{R}^m$ and $f'(x)$ is an $m \times n$ matrix with rows $f_i'(x)$, $i = 1, \ldots, m$, we obtain

$$u(x) = -(f'(x)(f_0'(x))^*)^{-1}[-f(x) + f'(x)(f_0'(x))^*] \tag{2.49}$$

$$p(x) = -[(f_0'(x))^* + (f'(x))^* u(x)]. \tag{2.50}$$

Substituting (2.49) in (2.50), we obtain

$$p(x) = [(I_n - \Pi(x))(f_0'(x))^* + (f'(x))^*(f'(x)(f'(x))^*)^{-1} f(x)] \tag{2.51}$$

where

$$\Pi(x) = (f'(x))^*(f'(x)(f'(x))^*)^{-1} f'(x). \tag{2.52}$$

According to Section 1.3.4, all the above calculations are valid provided the gradients $f_i'(x)$, $i = 1, \ldots, m$ are linearly independent, which we implicitly assumed earlier.

Formulae (2.49) and (2.50) may be used to calculate $u(x)$ and $p(x)$ explicitly. However, a computationally more economical approach involves direct solution of problem (2.48) by transferring to the dual problem:

$$\max_u\{-1/2\|f_0'(x) + \sum_{i=1}^m u^i f_i'(x)\|^2 + \sum_{i=1}^m u^i f_i(x)\}. \tag{2.53}$$

Here, since (2.48) involves only equality-type constraints, there are no constraints on the signs of the variables and direct solution is possible using the conjugate gradient method (Section 1.3.1).

Thus, in this case, the linearization algorithm takes the following form:

0. Suppose that N is a sufficiently large number and that x_k has already been constructed.

1. We solve problem (2.53) for $x = x_k$ and find a coefficient $u_k = u(x_k)$. Substituting this in (2.50), we find $p_k = p(x_k)$.

2. We divide $\alpha = 1$ by two until the inequality

$$f_0(x_k + \alpha_k p_k) + NF(x_k + \alpha_k p_k) \leq f_0(x_k) + NF(x_k) - \epsilon\alpha_k\|p_k\|^2$$

is satisfied, in order to find α_k.

3. We set $x_{k+1} = x_k + \alpha_k p_k$ and return to 1.

4. Stopping criterion: $\|p_k\| = 0$.

2.1.8 Simple Constraints

Suppose now that the problem only has constraints on the variations of the individual coordinates, i.e. the problem is of the form

$$\min_x\{f_0(x) | a^j \leq x^j \leq b^j,\ j = 1, \ldots n\} \tag{2.54}$$

where the values $a^j = -\infty$ and $b^j = +\infty$ are not ruled out.

For this problem, it is sensible to use all the constraints in the auxiliary problem. Here

$$F(x) = \max\{0, a^1 - x^1, \ldots, a^n - x^n, x^1 - b^1, \ldots, x^n - b^n\}$$

and the auxiliary problem takes the form

$$\min\{\sum_{j=1}^n \frac{\partial f_0}{\partial x^j} p^j + \frac{1}{2}\sum_{j=1}^n (p^j)^2 | a^j \leq x^j + p^j \leq b^j,\ j = 1, \ldots, n\}. \tag{2.55}$$

Since the goal function is now a sum of terms, each of which depends on its own variables, which vary independently of those of the other terms, the problem is easy to solve:

$$p^j(x) = \begin{cases} a^j - x^j & \text{if } -\frac{\partial f_0}{\partial x^j} \le a^j - x^j \\ -\frac{\partial f_0}{\partial x^j} & \text{if } a^j - x^j \le \frac{\partial f_0}{\partial x^j} \le b^j - x^j \\ b^j - x^j & \text{if } -\frac{\partial f_0}{\partial x^j} \ge b^j - x^j \end{cases} \tag{2.56}$$

If $p^j(x)$ is known, it is easy to find Kuhn–Tucker coefficients. We then denote the coefficients corresponding to the inequalities

$$x^j + p^j - b^j \le 0, \quad -x^j - p^j + a^j \le 0$$

by u_-^j and u_+^j, respectively. Then the Lagrange function takes the form

$$\sum_{j=1}^{n} \left[\frac{1}{2}(p^j)^2 + \frac{\partial f_0}{\partial x^j} p^j + u_+^j(a^j - x^j - p^j) + u_-^j(x^j + p^j - b^j) \right]$$

By definition, according to Theorem 2.6, the Kuhn–Tucker coefficients should have the following properties. Firstly, a minimum of the Lagrange function with respect to p should be attained by the solution $p(x)$. Whence, by differentiation, we obtain

$$\frac{\partial f_0}{\partial x^j} - u_+^j + u_-^j + p^j = 0, \quad j = 1, \ldots, n.$$

Secondly,

$$\begin{aligned} u_+^j &\ge 0, \quad u_+^j(a^j - x^j - p^j) = 0 \\ u_-^j &\ge 0, \quad u_-^j(x^j + p^j - b^j) = 0, \quad j = 1, \ldots, n. \end{aligned}$$

Since $p = p(x)$, it is then known that it is easy to determine u_+^j and u_-^j from equation (2.56):

$$\begin{aligned} u_-^j(x) &= 0, \quad u_+^j(x) = \frac{\partial f}{\partial x^i} + p^j(x) = \frac{\partial f}{\partial x^j} + a^j - x^j, \text{ if } -\frac{\partial f}{\partial x^j} \le a^j - x^j \\ u_-^j(x) &= u_+^j(x) = 0, \text{ if } a^j - x^j < -\frac{\partial f}{\partial x^j} < b^j - x^j \\ u_+^j(x) &= 0, \quad u_-^j(x) = -\frac{\partial f}{\partial x^j} - p^j(x) = -\frac{\partial f}{\partial x^j} - b^j + x^j \text{ if } -\frac{\partial f_0}{\partial x^j} \ge b^j - x^j. \end{aligned}$$

Thus, in the case of simple constraints, the auxiliary problem is easy to solve and Kuhn–Tucker coefficients are easy to determine.

However, if the initial point x_0 satisfies the constraint $a \le x_0 \le b$, since by construction $a \le x + p(x) \le b$, then $a \le x + \alpha p(x) \le b$, $0 \le \alpha \le 1$ and thus $F(x + \alpha p(x)) = 0$, $0 \le \alpha \le 1$.

Whence it follows that when choosing α at each step of the linearization method according to formula (2.5), there is no need to know the constant N, and thus no need to calculate Kuhn–Tucker coefficients.

Thus, for simple constraints, the algorithm has the following form:

0. A point x_0 is chosen so that $a^j \le x_0^j \le b^j$, $j = 1, \dots, n$.

1. If the point x_k has already been constructed, the vector $p_k = p(x_k)$ is calculated using formula (2.56).

2. We divide $\alpha = 1$ by two until the inequality

$$f_0(x_k + \alpha p_k) \le f_0(x_k) - \epsilon\alpha \|p_k\|^2$$

   is satisfied. We set α_k to equal the α' thus obtained and $x_{k+1} = x_k + \alpha_k p_k$. The algorithm returns to 1.

3. Stopping criterion: $p_k = 0$.

Theorem 2.3 *If the numbers a^j, b^j, $j = 1, \dots, n$ are finite and the gradient of the function f_0 satisfies a Lipschitz condition then the necessary conditions for a minimum are satisfied at all the limit points of the sequence x_k, $k = 0, 1, \dots$.*

Proof. The proof follows directly from Theorem 2.1. Clearly, in this case, the sequence x_k does not go outside a bounded region, the auxiliary problem (2.55) is always solvable and the coefficients u_+^j and u_-^j are bounded, as can be seen directly from the above formulae. Thus, all the assumptions of Theorem 2.1 are valid. □

2.1.9 Choice of Parameters in the Linearization Method. Modified Algorithm

As can be seen from Sections 2.1.1–2.1.3, for the method to converge, we must choose the parameters N and δ; generally speaking, these are a priori unknown. Thus, we require well-reasoned criteria for selecting or changing these parameters.

As previously mentioned, computational experience shows that it is best to take the parameter δ to be large so that all the constraints are taken into account from the start. Thus, we shall consider only the case in which $\delta = +\infty$.

Then the auxiliary problem will have the form

$$\min\{(f_0'(x), p) + 1/2\|p\|^2 | (f_i'(x), p) + f_i(x) \le 0,\ i \in I,\ x + p \in M\} \quad (2.57)$$

where M is a convex set with a simple structure.

We note one important feature of the auxiliary problem, namely that its resolution, which involves determining the direction of the displacement at each iteration, does not depend on the choice of N. The value of N only affects the choice of the step α.

The linearization method may be interpreted as a method of minimizing a penalty function

$$\Phi_N(x) = f_0(x) + N \max\{0, f_i(x) | i \in I\}$$

on the set M, since this function decreases at each step. It may seem that if the linearization method converges then it should converge to a point x_* which is a minimum of $\Phi_N(x)$ over M. A simple example shows that this is not the case.

In fact, let $f_0(x) = -x^2$, $f_1(x) = -x$ and $f_2(x) = x - 1$, $M = \mathbb{R}^n$; in other words, solve the problem

$$\min\{-x^2 | -x \leq 0, x - 1 \leq 0\} = \min\{-x^2 | 0 \leq x \leq 1, -\infty \leq x \leq +\infty\}$$

Clearly, this has solution $x_* = 1$. The linearization method immediately leads to a solution if it begins from any point $x > 0$, $x \leq 1$. At the same time, the penalty function

$$\Phi_N(x) = -x^2 + N \max\{0, -x, x - 1\}$$

has lower bound equal to $-\infty$ for all $N \geq 0$.

Thus, it is impossible to solve the original problem by the penalty function method. It is easy to check that at the point $x_* = 1$, the necessary conditions for a minimum of $\Phi_N(x)$ are satisfied; however this is not a global minimum, but a local minimum.

From the above, it follows that the value of N for non-convex problems may be in no way associated with the penalty function method. On the other hand, if all the functions involved in the problem are convex and N is chosen so that the conditions of the main assumptions of Section 2.1.1 are satisfied, then

$$N > \sum_{j \in I} u_*^j$$

where u_* is the Kuhn–Tucker vector of problem (2.2) and at the same time u_* is also the Kuhn–Tucker vector of the auxiliary problem (2.4) for $x = x_*$. Thus, from Corollary 1 of Theorem 2.14, the solution of problem (2.2) is also a minimum of the penalty function $\Phi_N(x)$. Therefore, in the convex case, the linearization method may also be viewed as a modification of the penalty function method.

We now describe an algorithm for changing N, which, with reasonable assumptions, guarantees the convergence of the linearization method.

Modified Linearization Algorithm Suppose that a point x_0 and a number $N_0 > 0$ are chosen. We set $C_0 = f_0(x) + N_0 F(x_0)$. If x_k, N_k and C_k have already been constructed then subsequent values of these may be assigned according to the following rule:

1. We solve the auxiliary problem (2.57) for $x = x_k$ and determine $p_k = p(x_k)$ and the Lagrange multipliers $u_k^i = u^i(x_k)$, $i \in I$.

2. If $N_k > \sum_{i \in I} u_k^i$, then

$$N_{k+1} = N_k, \quad C_{k+1} = C_k, \quad x_{k+1} = x_k + \alpha_k p_k,$$

where α_k is chosen by dividing $\alpha = 1$ by two until inequality (2.5) is satisfied with $N = N_{k+1}$.

3. If $N_k \leq \sum_{i\in I} u_k^i$, then

$$N_{k+1} = 2\sum_{i\in I} u_k^i$$

4. If $f_0(x_k) + N_{k+1}F(x_k) \leq C_k$, then

$$C_{k+1} = C_k, \quad x_{k+1} = x_k + \alpha_k p_k.$$

where α_k is chosen as before.

5. If $f(x_k) + N_{k+1}F(x_k) > C_k$, then

$$C_{k+1} = f(x_k) + N_{k+1}F(x_k), \quad x_{k+1} = x_0.$$

Theorem 2.4 *Suppose that the following conditions are satisfied:*

a) in any compact region $f_i'(x)$, $i \in \{0\} \cup I$ satisfy a Lipschitz condition with a constant which may depend on the region;

b) in any compact region the auxiliary problem (2.57) is solvable and the sum of the Lagrange multipliers is bounded by a constant which depends on the region;

c) $\inf_x\{f_0(x)|x \in M\} = \mu > -\infty$;

d) the set

$$W_\alpha = \{x|F(x) \leq \alpha, \ x \in M\}$$

is bounded.

Then the modified linearization method converges in the sense that any limit point of the sequence $\{x_k\}$ it generates satisfies the constraints

$$f_i(x) \leq 0, \quad i \in I, \quad x \in M$$

and the necessary conditions for a minimum are satisfied there.

Proof. We note that, by the construction of the algorithm, N_k may only increase. For each increase $N_{k+1} \geq 2N_k$, so that if there are infinitely many increases $N_k \to +\infty$. Moreover, by construction, $x_k \in M$ for all k.

We set

$$\alpha = F(x_0) + \frac{f(x_0) - \mu}{N_0} \tag{2.58}$$

and suppose that N_α is the constant which, according to assumption b) bounds the sum of the Lagrange multipliers of the auxiliary problem in the region W_α.

The whole of the sequence $\{x_k\}$ generated by the algorithm may be divided into segments, with initial point x_0 which return to x_0 after finitely many points.

To be definite, we consider the first of these segments $\{x_k\}_{k=0}^{l}$, such that $x_{l+1} = x_0$. Within this segment $C_k = C_0$.

We shall show that for all $k = 0, 1, \ldots, l$ the inequality

$$f_0(x_k) + N_k F(x_k) \leq C_0 \tag{2.59}$$

is satisfied. Clearly, if this inequality is satisfied for a given k, then for $k+1$, by virtue of stages 2–4 of the algorithm, independently of the relationship between N_{k+1} and N_k, we have

$$f(x_{k+1}) + N_{k+1}F(x_{k+1}) < f_0(x_k) + N_{k+1}F(x_k) \leq C_k = C_0.$$

It follows from (2.59) that

$$F(x_k) \leq \frac{C_0 - f_0(x_k)}{N_k} \leq \frac{C_0 - \mu}{N_0} = \alpha$$

since $N_k \geq N_0$, $f(x_k) \geq \mu$.

Thus, $x_k \in W_\alpha$ for all k in the chosen segment, whence for all k, since each of the chosen segments of the sequence $\{x_k\}$ begins with x_0.

But, from the fact that $x_k \in W_\alpha$ and by virtue of assumptions b) and d), it follows that, after a finite number of increases in N_k, this value becomes greater than N_α and stops changing. From this instant, the algorithm begins to work like the linearization method with a fixed constant N; whence its convergence follows from Theorem 2.1. □

We note that the sequence $\{x_k\}$ should return to the point x_0 quite rarely since even the first estimate of the value of N obtained at the point x_0 gives a good approximation. Moreover, if $F(x_k) = 0$ (and, in practice, if $F(x_k)$ is small enough), then, according to stage 4 of the algorithm x_0 should not be returned to. Clearly, this explains the fact that in practical calculations it was simply sufficient to increase N each time at stage 3 without returning to the initial point.

The only condition of Theorem 2.4 which is difficult to verify is condition b). But when one properly assesses what the theorem actually states this condition does not seem at all inflexible. The solvability of systems of nonlinear inequalities follows directly from the theorem. It is sensible to record this fact.

Theorem 2.5 *Under the conditions of Theorem 2.4 the system of inequalities*

$$f_i(x) \leq 0, \quad i \in I, \quad x \in M$$

has a solution.

We now describe two important cases in which the conditions of Theorem 2.4 are satisfied. To avoid repetition, in what follows, we shall assume that condition a) of Theorem 2.4 is satisfied.

Theorem 2.6 *Suppose that in any compact region the gradients $f_i'(x)$ of the functions $f_i(x)$, $i = 1, \ldots, m$ are linearly independent, that $f_0(x) \geq \mu > -\infty$ for all $x \in \mathbb{R}^n$ and that the set*

$$W_\alpha = \{x | F(x) = \max_{i=1,\dots,m} |f_i(x)| \le \alpha\}$$

is bounded. Then there exists a solution of the problem

$$\min_x \{f_0(x) | f_i(x) = 0, \; i = 1, \dots, m\},$$

and the modified linearization algorithm generates a sequence at each limit point of which the constraints of the problem and the necessary conditions for an extremum are satisfied.

Proof. As shown in Section 2.1.7, for this problem, the solution of the auxiliary problem and the Lagrange multipliers are given by formulae (2.49) and (2.50). By virtue of the assumptions of Theorem 2.6, the matrix $[f'(x)(f'(x))^*]^{-1}$ exists and is continuous, whence it is bounded on a compact region. Formulae (2.49) show that the vector $u(x)$ is also bounded on any compact region. Thus, all the conditions of Theorem 2.4 are satisfied and the result follows. □

We now turn to the convex case.

Theorem 2.7 *Suppose that the functions $f_i(x)$, $i \in I$ and the set M are convex and that there exists a point $\overline{x} \in M$ such that*

$$f_i(\overline{x}) \le -\gamma < 0, \quad i \in I.$$

Then the following bound holds for the Lagrange multipliers of the auxiliary problem (2.57):

$$\sum_{i \in I} u^i(x) \le \frac{1}{2\gamma} [\|f_0'(x)\| + \|\overline{x} - x\|]^2$$

Proof. Suppose that $p(x)$ and $u^i(x)$ are, respectively, the solution and the Lagrange multipliers of problem (2.57). Then

$$\begin{aligned} (f_0'(x), p(x)) + 1/2\|p(x)\|^2 &\le L(p, u(x)) \\ &= (f_0'(x), p) + 1/2\|p\|^2 \\ &\quad + \sum_{i \in I} u^i(x)[(f_i'(x), p) + f_i(x)], \quad x + p \in M. \end{aligned}$$

In addition, since the functions f_i are convex,

$$\begin{aligned} (f_i'(x), p) + f_i(x) &\le f_i(x + p), \quad i \in I \\ (f_0'(x), p(x)) + 1/2\|p(x)\|^2 &\ge \min_p [(f_0'(x), p) + 1/2\|p\|^2] \\ &= -1/2\|f_0'(x)\|^2. \end{aligned}$$

Whence, and by virtue of the previous inequality, for $\overline{p} = \overline{x} - x$, we obtain

$$\begin{aligned} -1/2\|f_0(x)\|^2 &\le (f_0'(x), p) + 1/2\|\overline{p}\|^2 + \sum_{i \in I} u^i(x) f_i(\overline{x}) \\ &\le \|f_0'(x)\|\|\overline{p}\| + 1/2\|\overline{p}\|^2 - \gamma \sum_{i \in I} u^i(x) \end{aligned}$$

or, after simple transformations

$$\sum_{i\in I} u^i(x) \leq \frac{1}{2\gamma}[\|\|f_0'(x)\| + \|\overline{p}\|\|]^2,$$

which completes the proof of the theorem. □

Comparing this result with Theorem 2.4, we derive the following conclusion.

Theorem 2.8 *Suppose that the following conditions are satisfied:*

a) The functions $f_i(x)$, $i \in I$ and the set M are convex and there exists a point $\overline{x} \in M$ such that

$$f_i(\overline{x}) \leq -\gamma < 0, \quad i \in I.$$

b) The set

$$W_\alpha = \{x | f_i(x) \leq \alpha, \; i \in I, \; x \in M\}$$

is bounded.

c) $\inf_x \{f_0(x) | x \in M\} = \mu > -\infty$.

Then the modified algorithm converges in the sense described in Theorem 2.4.

We note that we did not assume that the function $f_0(x)$ is convex. If we were to assume this, any limit point of the sequence generated by the algorithm would be a solution of the convex programming problem

$$\min_x \{f_0(x) | f_i(x) \leq 0, \; i \in I, \; x \subset M\}.$$

2.2 Resolution of Systems of Equations and Inequalities

Clearly, the resolution of systems of nonlinear equations and inequalities is a necessary element of mathematical programming. The algorithms of the previous section are naturally suitable for this. Clearly, if the function to be minimized is assumed to be identically zero then any point satisfying the constraints will be a solution of the mathematical programming problem.

However, as we shall see below, the problem of finding a solution of a system of equations and inequalities has a number of features which enable us to guarantee a far higher rate of convergence than that which can be achieved in general optimization problems without additional effort.

Therefore, let us consider the system

$$f_i(x) \leq 0, \quad i \in I^-, \quad f_i(x) = 0, \quad i \in I^0 \tag{2.60}$$

where I^- and I^0 are finite sets of indices. As in the previous section, without loss of generality, in what follows, we shall consider only the system of inequalities

$$f_i(x) \le 0, \quad i \in I = \{1, \ldots, m\}. \tag{2.61}$$

Resolution of (2.60) reduces to resolution of (2.61) using the techniques described in Section 2.1.

We set

$$\begin{aligned} F(x) &= \max\{0, f_1(x), \ldots, f_m(x)\} \\ I_\delta(x) &= \{i \in I | f_i(x) \ge F(x) - \delta\}, \quad \delta \ge 0. \end{aligned} \tag{2.62}$$

Throughout this section, we shall assume that the gradients $f_i'(x)$ satisfy a Lipschitz condition in any compact region.

2.2.1 The Auxiliary Problem

We associate each point x with the auxiliary quadratic programming problem

$$\min_p \{1/2\|p\|^2 | (f_i'(x), p) + f_i(x) \le 0, \; i \in I_\delta(x)\}. \tag{2.63}$$

We denote the solution of this problem by $p(x)$ and the corresponding Lagrange multipliers by $u^i(x) \ge 0$, $i \in I_\delta(x)$. According to the theory of necessary conditions for extrema, discussed in Section 1.2, the following conditions should be satisfied:

$$\begin{aligned} (p(x))^* + \sum_{i \in I_\delta(x)} u^i(x) f_i'(x) &= 0 \\ u^i(x)[(f_i'(x), p(x)) + f_i(x)] &= 0, \quad i \in I_\delta(x). \end{aligned} \tag{2.64}$$

But by scalar multiplication of the first of these equations by $p(x)$ and taking into account the second equation, we easily obtain

$$\|p(x)\|^2 = \sum_{i \in I_\delta(x)} u^i(x) f_i(x). \tag{2.65}$$

2.2.2 The Algorithm

We now develop the computational procedure which, subject to certain assumptions, will be used to solve the system (2.61).

Suppose that an initial point x_0 and ϵ, $0 < \epsilon < 1$ are already chosen. If the point x_k has already been constructed, then we solve problem (2.63) and obtain $p_k = p(x_k)$. We compute α_k by dividing $\alpha = 1$ by two until the inequality

$$F(x_k + \alpha_k p_k) \le (1 - \epsilon \alpha_k) F(x_k) \tag{2.66}$$

is satisfied. We set

$$x_{k+1} = x_k + \alpha_k p_k.$$

The iteration terminates.

In what follows we shall describe the conditions under which this algorithm converges.

2.2.3 Convergence of the Algorithm

We shall give a number of bounds. Suppose that $i \in I_\delta(x)$ and that problem (2.63) is solvable. Then, for $p = p(x)$, using the mean value formula, we obtain

$$\begin{aligned} f_i(x+\alpha p) &= f_i(x) + \alpha(f_i'(x+\theta\alpha p), p) \\ &= f_i(x) + \alpha(f_i'(x), p) + \alpha(f_i'(x+\theta\alpha p) - f_i'(x), p) \\ &\leq f_i(x) - \alpha f_i(x) + \alpha^2 L\|p\|^2 \end{aligned}$$

where $0 \leq \theta \leq 1$ and L is the Lipschitz constant for the gradients $f_i'(x)$. In addition, we used the fact that the vector p satisfies the constraints of problem (2.63).

This formula may be rewritten in the form

$$f_i(x+\alpha p) \leq (1-\alpha)F(x) + \alpha^2 L\|p\|^2, \quad i \in I_\delta(x) \tag{2.67}$$

where $0 \leq \alpha \leq 1$.

If $i \notin I_\delta(x)$, then

$$\begin{aligned} f_i(x+\alpha p) &= f_i(x) + \alpha(f_i'(x+\theta p), p) \\ &\leq f_i(x) + \alpha K\|p\| \\ &\leq F(x) - \delta + \alpha K\|p\| \end{aligned} \tag{2.68}$$

where K is a constant which bounds the norm of the gradient in a sufficiently broad region. Moreover,

$$(1-\alpha)F(x) \geq F(x) - \delta + \alpha K\|p\|$$

if

$$\alpha \leq \alpha_1 = \frac{\delta}{F(x) + K\|p\|}. \tag{2.69}$$

Thus, for $\alpha \leq \min[1, \alpha_1]$, from (2.67) and (2.68), we deduce the following inequality

$$f_i(x+\alpha p) \leq (1-\alpha)F(x) + \alpha^2 L\|p\|^2$$

which is valid for all $i \in I$, so that

$$F(x+\alpha p) \leq (1-\alpha)F(x) + \alpha^2 L\|p\|^2, \tag{2.70}$$

where $\alpha \leq \min[1, \alpha_1]$.

Based on the bound (2.70), we introduce a number of criteria for convergence.

Theorem 2.9 *Suppose that the set $\Omega_0 = \{x | F(x) \leq F(x_0)\}$ is compact and that the auxiliary problem (2.63) has a uniformly bounded solution in Ω_0. Then the algorithm of Section 2.2.2 generates a sequence x_k for which $F(x_k) \to 0$.*

Proof. By assumption, for some constant C,

$$\|p(x)\| \le C, \quad x \in \Omega_0.$$

The bound (2.70) at the point $x = x_k$ may then be rewritten in the form

$$\begin{aligned} F(x_k + \alpha p_k) &\le (1-\alpha)F(x_k) + \alpha^2 LC^2 \\ &= F(x_k) - \alpha F(x_k)\left[1 - \alpha\frac{LC^2}{F(x_k)}\right] \end{aligned}$$

If

$$1 - \alpha\frac{LC^2}{F(x_k)} \ge \epsilon,$$

in other words, if

$$\alpha \le \frac{1-\epsilon}{LC^2}F(x_k)$$

then

$$F(x_k + \alpha p_k) \le (1 - \alpha\epsilon)F(x_k). \tag{2.71}$$

We note that inequality (2.71) is valid for

$$\alpha \le \min\left[1, \frac{\delta}{F(x_k) + K\|p_k\|}, \frac{1-\epsilon}{LC^2}F(x_k)\right]$$

We then recall that α_k was chosen by dividing unity by two until inequality (2.66) was satisfied. Thus,

$$\alpha_k \ge \frac{1}{2}\min\left[1, \frac{\delta}{F(x_k) + K\|p_k\|}, \frac{1-\epsilon}{LC^2}F(x_k)\right]. \tag{2.72}$$

By virtue of (2.71), $F(x_k)$ is monotonic decreasing; thus, if $F(x_k)$ does not converge to zero then $F(x_k) \ge v > 0$ for all k.

Using (2.72) and evident coarse forms of this inequality, we obtain

$$\alpha_k \ge \frac{1}{2}\min\left[1, \frac{\delta}{F(x_0) + KC}, \frac{1-\epsilon}{LC^2}v\right] = \overline{\alpha} > 0.$$

Thus,

$$F(x_k + \alpha_k p_k) = F(x_{k+1}) \le (1 - \overline{\alpha}\epsilon)F(x_k);$$

whence, it follows that $F(x_k) \to 0$. This completes the proof of the theorem. □

Remark. Since $F(x_k) \to 0$, the inequality

$$\alpha_k \ge \frac{1}{2}\frac{1-\epsilon}{LC^2}F(x_k)$$

follows from (2.72). Substituting the right-hand side of this inequality in (2.71) instead of α, we obtain the coarser inequality

$$\begin{aligned} F(x_{k+1}) &\leq (1-\alpha_k\epsilon)F(x_k) \\ &\leq \left(1-\frac{1-\epsilon}{2LC^2}F(x_k)\right)F(x_k). \end{aligned}$$

It is shown in [28, p.205] that the above inequality implies that there exists a constant $C_1 > 0$ such that

$$F(x_k) \leq C_1/k.$$

Thus, the convergence may be quite slow.

Of course, this estimate of a slow rate of convergence is not a result of the fact that the convergence is actually poor, but follows from the coarseness of the original assumption that $p(x)$ is simply bounded. Under more precise assumptions, which guarantee that $p(x)$ tends to zero together with $F(x)$, the bound obtained is more precise.

Theorem 2.10 *If the set $\Omega_0 = \{x|F(x) \leq F(x_0)\}$ is compact, and the auxiliary problem (2.63) on this set is solvable with uniformly bounded Lagrange multipliers, then $F(x_k) \to 0$ and the inequality*

$$F(x_{k+1}) \leq qF(x_k), \quad 0 < q < 1$$

holds.

Proof. Suppose that for all $x \in \Omega_0$

$$\sum_{i\in I_\delta(x)} u^i(x) \leq N.$$

Then it follows from (2.65) that

$$\|p(x)\|^2 \leq NF(x) \tag{2.73}$$

whence inequality (2.70) then gives the bound

$$\begin{aligned} F(x_k+\alpha p_k) &\leq (1-\alpha)F(x_k)+\alpha^2 LNF(x_k) \\ &= [1-\alpha(1-\alpha LN)]F(x_k). \end{aligned}$$

Consequently, for

$$\alpha \leq (1-\epsilon)/(LN) \tag{2.74}$$

the inequality

$$F(x_k+\alpha p_k) \leq (1-\alpha\epsilon)F(x_k) \tag{2.75}$$

holds. Thus, if

$$\alpha \leq \min\left[1, \frac{\delta}{F(x_k)+K\|p_k\|}, \frac{1-\epsilon}{LN}\right] \tag{2.76}$$

then inequality (2.75) holds.

Taking into account the fact that α_k was chosen by dividing by two, we obtain

$$\begin{aligned} F(x_k + \alpha_k p_k) &\le (1 - \alpha_k \epsilon) F(x_k) \\ \alpha_k &\ge \frac{1}{2} \min \left[1, \frac{\delta}{F(x_k) + K\|p_k\|}, \frac{1-\epsilon}{LN} \right]. \end{aligned}$$

Then, it is clear that α_k remains greater than some value $\overline{\alpha}$, so that we have

$$F(x_{k+1}) \le qF(x_k), \quad q = 1 - \overline{\alpha}\epsilon.$$

We then note that since $F(x_k) \to 0$, it follows that $\|p_k\| \to 0$ (see (2.73)), whence, for large k

$$\frac{\delta}{F(x_k) + K\|p_k\|} \ge 1.$$

Thus, for large k,

$$\alpha_k \ge 1/2 \min[1, (1-\epsilon)/(LN)]$$

and we may set

$$\overline{\alpha} = 1/2 \min[1, (1-\epsilon)/(LN)].$$

Corollary *If the functions $f_i(x)$, $i \in I$ are convex, the set Ω is compact and there exists a point $\overline{x}$ such that*

$$f_i(\overline{x}) \le -\gamma < 0, \quad i \in I$$

then

$$F(x_{k+1}) \le qF(x_k), \quad 0 \le q < 1.$$

Proof. Clearly, by virtue of Theorem 2.7, for $f_0(x) = 0$

$$\sum_{i \in I} u^i(x) \le \frac{1}{2\gamma} \|x - \overline{x}\|^2.$$

Thus, all the conditions of Theorem 2.10 are satisfied. □

We now show that with fairly natural assumptions a stronger assertion may be made.

Theorem 2.11 *Suppose that the conditions of Theorem 2.10 are satisfied and that, in addition, for all x satisfying the inequality $F(x) \le w$ for some $w > 0$, there exists a constant C such that*

$$\|p(x)\| \le CF(x). \tag{2.77}$$

Then the sequence x_k converges to some point x_* which is a solution of the system (2.61), and for k sufficiently large

$$F(x_{k+1}) \leq LC^2F^2(x_k). \tag{2.78}$$

Proof. According to the previous theorem. $F(x_k) \to 0$. Thus, for large k, we have $F(x_k) \leq w$. Moreover, we may also assume that

$$\frac{\delta}{F(x_k) + K\|p_k\|} \geq 1,$$

whence, inequality (2.70) together with (2.77) show that

$$\begin{aligned} F(x_k + \alpha p_k) &\leq (1-\alpha)F(x_k) + \alpha^2 L\|p_k\|^2 \\ &\leq (1-\alpha)F(x_k) + \alpha^2 LC^2F^2(x_k) \end{aligned} \tag{2.79}$$

for $0 \leq \alpha \leq 1$ and k sufficiently large. Whence

$$F(x_k + \alpha p_k) \leq [1 - \alpha(1 - \alpha LC^2F(x_k))]F(x_k).$$

If

$$1 - \alpha LC^2F(x_k) \geq \epsilon,$$

i.e. if

$$\alpha \leq \frac{1-\epsilon}{LC^2F(x_k)}$$

then

$$F(x_k + \alpha p_k) \leq (1 - \alpha\epsilon)F(x_k) \tag{2.80}$$

for

$$\alpha \leq \min\left[1, \frac{1-\epsilon}{LC^2F(x_k)}\right].$$

But for k sufficiently large

$$\frac{1-\epsilon}{LC^2F(x_k)} \geq 1$$

whence, because of the rule for choosing α_k, this value becomes equal to one.

Thus, α_k is chosen equal to 1 for k sufficiently large. Equations (2.79) and (2.80) then take on the form

$$F(x_{k+1}) \leq LC^2F^2(x_k) \tag{2.81}$$

$$F(x_{k+1}) \leq (1-\epsilon)F(x_{k_1}) \tag{2.82}$$

These inequalities are satisfied for $k \geq k_1$, where k_1 is sufficiently large. It now follows from (2.82) that

$$F(x_k) \leq (1-\epsilon)^{k-k_1}F(x_{k_1}).$$

Comparing this with (2.77), we obtain

$$\|p_k\| \le C(1-\epsilon)^{k-k_1} F(x_{k_1}).$$

Consequently, the series $x_{k_1} + p_{k_1} + \ldots + p_{k-1} = x_k$ converges as $k \to \infty$, since it is bounded above by a convergent geometric progression.

Therefore,

$$\begin{aligned} x_k &\to x_* \\ F(x_*) &= \lim_{k\to\infty} F(x_k) = 0, \end{aligned}$$

which completes the proof of the theorem. □

We now determine a number of conditions under which inequality (2.77) holds. These conditions show that satisfaction of inequality (2.77) is quite natural and is characterized by a certain nondegeneracy of the system of inequalities (2.61).

Theorem 2.12 *Suppose that the set $\Omega_0 = \{x|F(x) \le F(x_0)\}$ is compact and that the auxiliary problem is solvable for $x \in \Omega_0$ with uniformly bounded Lagrange multipliers. Suppose also that there exist $\omega > 0$ and $\delta_0 > 0$ such that whenever x satisfies the inequality $0 < F(x) \le \omega$, the value*

$$l(x,\delta_0) = \min_{\lambda}\{\| \sum_{i\in I_{\delta_0}(x)} \lambda_i f_i'(x)\| | \lambda_i \ge 0, \sum_{i\in I_{\delta_0}(x)} \lambda_i = 1\}$$

is bounded below by a number $\gamma > 0$. Then the algorithm for solving the system of inequalities converges, i.e. $F(x_k) \to 0$, $x_k \to x_*$, $F(x_*) = 0$ and

$$F(x_{k+1}) \le C_1 F^2(x_k)$$

for k sufficiently large.

Proof. Based on Theorems 2.10 and 2.11, it is clear that it is sufficient to show that the assumptions of Theorem 2.12 ensure that inequality (2.77) is satisfied for points x_k with sufficiently large indices.

Since all the assumptions of Theorem 2.10 are satisfied $F(x_k)$ and $p(x_k)$ converge to zero. Let us consider sufficiently large indices k such that

$$F(x_k) \le \delta_0/2, \quad K\|p(x_k)\| \le \delta_0/2.$$

Suppose that $i \in I_\delta(x_k)$ and that $u^i(x_k) > 0$, i.e. an active constraint corresponding to the index i in the auxiliary problem:

$$(f_i'(x_k), p(x_k)) + f_i(x_k) = 0.$$

Whence,

$$\begin{aligned} |f_i(x_k)| &= |(f_i'(x_k), p(x_k))| \le K\|p(x_k)\| \le \delta_0/2 \\ F(x_k) - f_i(x_k) &\le \delta_0/2 - f_i(x_k) \le \delta_0 \end{aligned}$$

i.e. $i \in I_{\delta_0}(x_k)$.

Therefore, for large k, all active indices in the auxiliary problem belong to $I_{\delta_0}(x_k)$. Using (2.64), we then obtain

$$\begin{aligned}\|p(x_k)\| &= \|\sum_{i\in I_{\delta_0}(x_k)} u^i(x_k) f_i'(x_k)\| \\ &= \left(\sum_{i\in I_{\delta_0}(x_k)} u^i(x_k)\right) \|\sum_{i\in I_{\delta_0}(x_k)} \lambda_i f_i'(x_k)\| \\ &\geq \left(\sum_{i\in I_{\delta_0}(x_k)} u^i(x_k)\right) l(x_k, \delta_0)\end{aligned}$$

where we have introduced the notation

$$\lambda_i = \frac{u^i(x_k)}{\sum_{i\in I_{\delta_0}(x_k)} u^i(x_k)}.$$

Finally, we may write

$$\sum_{i\in I_{\delta_0}(x_k)} u^i(x_k) \leq \frac{\|p(x_k)\|}{l(x_k,\delta_0)} \leq \frac{1}{\gamma}\|p(x_k)\|.$$

Comparing this with (2.65), we obtain the inequality

$$\begin{aligned}\|p(x_k)\|^2 &\leq \left(\sum_{i\in I_{\delta_0}(x_k)} u^i(x_k)\right) F(x_k) \\ &\leq \frac{F(x_k)}{\gamma}\|p(x_k)\|\end{aligned}$$

or

$$\|p(x_k)\| \leq (1/\gamma)F(x_k).$$

Thus, inequality (2.77) is satisfied, whence all the conclusions of Theorem 2.11 hold. The proof is complete. □

Theorem 2.12 enables us to consider the case of convex inequalities in full.

Theorem 2.13 *Suppose that*

a) The $f_i(x)$, $i \in I$ are convex, continuously differentiable functions, the derivatives of which satisfy a Lipschitz condition;

b) The set $\Omega_0 = \{x|F(x) \leq F(x_0)\}$ is compact and there exists a point $\overline{x}$ such that

$$f_i(\overline{x}) \leq -\gamma < 0, \quad i \in I.$$

Then the algorithm generates a sequence x_k which converges to a solution of the system of inequalities with a quadratic rate of convergence:

$$F(x_{k+1}) \leq C_1 F^2(x_k).$$

Proof. Let $\delta_0 = \gamma/2$. Then, for $i \in I_{\delta_0}(x)$

$$f_i(x) \geq F(x) - \delta_0 \geq -\gamma/2,$$

i.e.

$$f_i(x) + \gamma/2 \geq 0, \quad i \in I_{\delta_0}(x). \tag{2.83}$$

Suppose that $F(x) > 0$, $p = \overline{x} - x$. Since $f_i(\overline{x}) \leq -\gamma$, $i \in I$, the norm of the vector $\overline{p}$ is always greater than some constant, because for x near to $\overline{x}$, $F(x) = 0$.

Since the functions $f_i(x)$ are convex

$$(f_i'(x), \overline{x} - x) + f_i(x) \leq f_i(\overline{x}) \leq -\gamma. \tag{2.84}$$

Thus,

$$(f_i'(x), \overline{p}) + f_i(x) + \gamma \leq 0$$

and, taking into account inequality (2.83), we obtain

$$(f_i'(x), \overline{p}) + \gamma/2 \leq 0, \quad i \in I_{\delta_0}(x). \tag{2.85}$$

We now choose arbitrary $\lambda_i \geq 0$ such that

$$\sum_{i \in I_{\delta_0}(x)} \lambda_i = 1.$$

Then, from (2.85), we obtain

$$\begin{aligned} \left(\sum_{i \in I_{\delta_0}(x)} \lambda_i f_i'(x), \overline{p} \right) + \gamma/2 &\leq 0 \\ -\| \sum_{i \in I_{\delta_0}(x)} \lambda_i f_i'(x) \| \|\overline{p}\| + \gamma/2 &\leq 0 \end{aligned}$$

and, since the λ_i are arbitrary,

$$-l(x, \delta_0) \|\overline{p}\| + \gamma/2 \leq 0.$$

Thus,

$$l(x, \delta_0) \geq \gamma/(2\|\overline{p}\|).$$

But this is once again the inequality which is required in Theorem 2.12. Thus, we have proved the theorem. □

2.3 Acceleration of the Convergence of the Linearization Method

The linearization method may also be applied to the case in which there are no constraints; in other words, when it is required to minimize a function $f_0(x)$ over the whole space $\mathbb{R}^n$. In this case, the auxiliary problem (2.4) takes the form

$$\min_p \{(f_0'(x), p) + 1/2\|p\|^2 | p \in \mathbb{R}^n\}.$$

The obvious solution of this is $p = -(f_0'(x))^*$. (We recall that f_0' is a row vector and that the superscript asterisk denotes transposition.)

From the above expression for the direction of the displacement p from the point x, it is clear that, in this case, the method coincides with the gradient descent method [28]. Thus, the rate of convergence of the linearization method as a whole cannot be greater than that of the gradient descent method.

In what follows, we analyse the rate of convergence of the linearization method. Based on this analysis, we shall construct a high-speed algorithm. It is also interesting to note that the presence of constraints may lead to an acceleration of the convergence and even to quadratic convergence in some cases, without special modifications to the algorithm.

The linearization method described in Section 2.1 provides for global convergence, i.e. converges from a relatively poor initial approximation. At the same time, the rate of convergence is bounded in a neighbourhood of the solution. Here, the high-speed algorithms operate in a neighbourhood of the solution, the true size of which is a priori unknown. This gives rise to the relatively complicated problem of combining global convergence and high speed using computational techniques which can automatically translate an algorithm providing global convergence into a local algorithm which rapidly reaches the solution. In what follows, we shall resolve this problem in a specific way. However, this approach is not entirely satisfactory from the computational point of view since the above method of translation requires additional processing.

2.3.1 Main Assumptions

We shall consider the minimization problem

$$\min_x \{f_0(x) | f_i(x) \leq 0,\ i \in I\},\quad I = \{1, 2, \ldots, m\}. \tag{2.86}$$

As in Section 2.1, we shall consider only inequality-type constraints, since equality-type constraints may be considered using obvious changes in the algorithm, as for example in Section 2.1. We recall that

$$\begin{aligned} F(x) &= \max\{0, f_1(x), \ldots, f_m(x)\} \\ I_\delta(x) &= \{i \in I | f_i(x) \geq F(x) - \delta\},\quad \delta > 0 \\ \Phi_N(x) &= f_0(x) + NF(x),\quad N \geq 0. \end{aligned} \tag{2.87}$$

Suppose that x_* is a solution of problem (2.86). According to the theorems of Section 1.2, the necessary conditions for an extremum are satisfied at this point

$$\begin{aligned} f_0'(x_*) + \sum_{i=1}^{m} u_*^i f_i'(x_*) &= 0 \\ u_*^i \geq 0, \ u_*^i f_i(x_*) &= 0, \ i = 1, \ldots, m. \end{aligned} \tag{2.88}$$

We denote

$$I_* = \{i \in I | f_i(x_*) = 0\}.$$

The set I_* is the set of indices of the active constraints in the solution. Throughout what follows, we shall suppose that the following assumptions are satisfied:

1. All the functions $f_i(x)$, $i = 0, 1, \ldots, m$ are three-times continuously differentiable.
2. The gradients $f_i'(x)$, $i \in I_*$ are linearly independent and $u_*^i > 0$, $i \in I_*$.
3. The condition
$$(L_{xx}''(x_*, u_*)p, p) > 0 \tag{2.89}$$
is satisfied for all $p \neq 0$ satisfying the equations
$$(f_i'(x_*), p) = 0, \quad i \in I_*. \tag{2.90}$$

By virtue of the results of Section 1.2, these assumptions are not too cumbersome and require only a certain regularity of the solution.

2.3.2 Local Analysis of the Auxiliary Problem

According to Section 2.1 the auxiliary problem at the point x has the form

$$\min_p \{(f_0'(x), p) + 1/2\|p\|^2 | (f_i'(x), p) + f_i(x) \leq 0, \ i \in I_\delta(x)\}. \tag{2.91}$$

Its solution and the Lagrange multipliers are denoted, respectively, by $p(x)$ and $u^i(x)$, $i \in I_\delta(x)$. We denote the set of active constraints of problem (2.91) by $I_\delta^0(x)$, i.e.

$$I_\delta^0(x) = \{i \in I_\delta(x) | (f_i'(x), p(x)) + f_i(x) = 0\}. \tag{2.92}$$

We then study the behaviour of the solution of problem (2.91) in a neighbourhood of the point x_*.

Lemma 2.2 *There exists a neighbourhood of the point x_* in which $I_* \subseteq I_\delta(x)$.*

Proof. It is easy to see that for such a neighbourhood we may take the set of points x satisfying the inequalities

$$F(x) < \delta/2, \ |f_i(x)| < \delta/2, \ i \in I_*.$$

This completes the proof. □

Lemma 2.3 *There exists a neighbourhood of the point x_* in which $p(x)$ and $u^i(x)$, $i \in I_*$ are differentiable with derivatives satisfying a Lipschitz condition and in this neighbourhood $I^0_\delta(x) = I_*$.*

Proof. By virtue of the necessary conditions for an extremum for problem (2.91) the equations

$$
\begin{aligned}
p(x) + (f_0'(x))^* + \sum_{i \in I_\delta(x)} u^i(x)(f_i'(x))^* &= 0 \qquad (2.93)\\
u^i(x) \geq 0,\ \ u^i(x)[(f_i'(x), p(x)) + f_i(x)] &= 0, \quad i \in I_\delta(x). \qquad (2.94)
\end{aligned}
$$

are satisfied. For $x = x_*$, $p(x_*) = 0$ and $u^i(x_*) = u^i_*$. In addition, it is clear that $I^0_\delta(x_*) = I_*$.

Let us consider the system of equations

$$
\begin{aligned}
\tilde{p} + (f_0'(x))^* + \sum_{i \in I_*} \tilde{u}^i (f_i'(x))^* &= 0\\
f_i'(x)\tilde{p} + f_i(x) &= 0, \quad i \in I_* \qquad (2.95)
\end{aligned}
$$

in the unknowns $\tilde{p}$ and $\tilde{u}^i$, $i \in I_*$. If we denote the matrix with rows $f_i'(x)$, $i \in I_*$, by $f'(x)$, the column vector with components $f_i(x)$, $i \in I_*$, by $f(x)$ and the column vector with components $\tilde{u}^i$, $i \in I_*$ by $\tilde{u}$, then the system of (2.95) may be rewritten in the form

$$
\begin{aligned}
\tilde{p} + (f'(x))^* \tilde{u} &= -(f_0'(x))^*\\
f'(x)p &= -f(x) \qquad (2.96)
\end{aligned}
$$

It is easy to write the solution of this system in explicit form by expressing p in terms of u from the first equation and substituting it in the second. After simple calculations, we obtain the following expressions:

$$
\begin{aligned}
\tilde{u}(x) &= (f'(x)f'^*(x))^{-1}[f(x) - f'(x)f_0'^*(x)] \qquad (2.97)\\
\tilde{p}(x) &= -(I - \Pi(x))f_0'^*(x) - f'^*(x)v(x) \qquad (2.98)
\end{aligned}
$$

where

$$
\begin{aligned}
\Pi(x) &= f'^*(x)[f'(x)f'^*(x)]^{-1}f'(x) \qquad (2.99)\\
v(x) &= (f'(x)f'^*(x))^{-1}f(x). \qquad (2.100)
\end{aligned}
$$

The inverse matrix which occurs in these expressions is well-defined since, by assumption, the vectors $f_i'(x_*)$, $i \in I_*$ are linearly independent, whence they are linearly independent in some neighbourhood of the point x_*.

We shall show that in some neighbourhood of the point x_*, the functions $\tilde{p}(x)$ and $\tilde{u}(x)$ are also solutions of the auxiliary problem (2.91). From formulae (2.97)–(2.100) and the differentiability properties of the f_i, it follows that $\tilde{p}(x)$ and $\tilde{u}(x)$ are twice continuously differentiable. In addition, these formulae show that the system (2.95) has a unique solution. Since for $x = x_*$ the system (2.95) has a solution $\tilde{p} = 0$, $\tilde{u} = u_*$, we have

$$\tilde{p}(x_*) = 0, \quad \tilde{u}(x_*) = u_*.$$

But, $u_* > 0$, by assumption, and thus, in some neighbourhood of x_*

$$\tilde{u}(x) \geq 0. \tag{2.101}$$

Furthermore, since $f_i(x_*) < 0$, $i \notin I_*$, we have

$$(f_i'(x), \tilde{p}(x)) + f_i(x) < 0, \quad i \notin I_* \tag{2.102}$$

in some neighbourhood of x_*.

We now set $\tilde{u}^i = 0$, $i \notin I_*$. Then, by virtue of (2.95), (2.101) and (2.102) we deduce that in a sufficiently small neighbourhood of x_* the vector $\tilde{p}(x)$ satisfies the constraints of the auxiliary problem and equations (2.93) and (2.94) are satisfied. But the last two equations are not only necessary but also sufficient equations for the vector $\tilde{p}(x)$ to be a solution of the auxiliary problem (2.91). In fact, we have actually shown that the vector $\tilde{p}(x)$ is a solution of the auxiliary problem (2.91) and that the $\tilde{u}(x)$ are the corresponding non-zero Lagrange multipliers. Here, by virtue of (2.95) and (2.102), the only active constraints in the auxiliary problem are those for which $i \in I_*$, i.e. $I_\delta^0(x) = I_*$. This completes the proof of the lemma. □

From this lemma it follows that when studying local properties of the vector $p(x)$ in a neighbourhood of the point x_* we may use the system (2.95) (or (2.96)), viewing it as a system in which p and u are implicit functions of x. Here, we may use the implicit function theorem [23], from which it follows that if the matrix of partial derivatives with respect to the desired variables p and u^i (Jacobian) on the left-hand side of (2.95) is nonsingular at $x = x_*$ then p and u^i are differentiable with respect to x and their derivatives may be found by differentiating (2.95) and solving the resulting system in the desired derivatives.

We now note that the system (2.95) is linear in p and u, thus the Jacobian at the point $x = x_*$ has the form

$$\begin{pmatrix} I_n & f'^*(x_*) \\ f'(x_*) & 0 \end{pmatrix}$$

In what follows, we shall show, for a more complicated case, that this matrix is nonsingular. Thus, as previously noted, in order to compute the matrix

$$p'(x_*) = \{\partial p^i / \partial x^j\}_{i,j=1,\ldots,n}$$

we need only differentiate formula (2.95) at the point $x = x_*$.

Carrying out this differentiation, taking into account the fact that $p(x_*) = 0$, we obtain

$$\begin{aligned} p'(x_*) + f_0''(x_*) + \sum_{i \in I_*} u_*^i f_i''(x_*) + \sum_{i \in I_*} f_i'^*(x_*)(u^i(x_*))' &= 0 \\ f'(x_*)p'(x_*) + f'(x_*) &= 0 \end{aligned}$$

or, in matrix form

$$\begin{aligned} p'(x_*) + L''_{xx}(x_*, u_*) + \sum_{i \in I_*} f_i'^*(x_*)(u^i(x_*))' &= 0 \\ f'(x_*)p'(x_*) + f'(x_*) &= 0. \end{aligned} \tag{2.103}$$

We now set $\Pi_* = \Pi(x_*)$. From (2.99) and the results of Section 1.3.5, it follows that Π_* is the projection operator onto the subspace orthogonal to the subspace $\{p | f'(x_*)p = 0\}$, where

$$\Pi_*^2 = \Pi_*, \quad \Pi_*^* = \Pi_*, \quad \Pi_* f'^*(x_*) = f'^*(x_*). \tag{2.104}$$

Multiplying the second equation of (2.103) by $f'^*(x_*)[f'(x_*)f'^*(x_*)]^{-1}$, we obtain

$$\Pi_* p'(x_*) + \Pi_* = 0. \tag{2.105}$$

If we now multiply the first equation of (2.103) by $I_n - \Pi_*$, by virtue of (2.104), we obtain

$$(I_n - \Pi_*)p'(x_*) + (I_n - \Pi_*)L''_{xx}(x_*, u_*) = 0. \tag{2.106}$$

Finally, adding (2.105) and (2.106), we obtain

$$p'(x_*) = -[\Pi_* + (I_n - \Pi_*)L''_{xx}(x_*, u_*)]. \tag{2.107}$$

Lemma 2.4 *The eigenvalues γ_j of the matrix $p'(x_*)$ may be characterized as follows: $\gamma_j = -1$, $j = 1, \ldots, |I_*|$, where $|I_*|$ is the number of indices in the set I_* of active constraints. The remaining $n - |I_*|$ eigenvalues coincide with the nonzero eigenvalues of the matrix $(I - \Pi_*)L''_{xx}(x_*, u_*)(I - \Pi_*)$, taken with the opposite sign.*

Proof. Suppose that σ is an eigenvalue and y is an eigenvector of the matrix $p'(x_*)$. Then according to formula (2.107)

$$-\Pi_* y - (I_n - \Pi_*)L''_{xx}(x_*, u_*)y = \sigma y = \sigma \Pi_* y + \sigma(I_n - \Pi_*)y$$

whence

$$\Pi_* y = -\sigma \Pi_* y, \quad (I_n - \Pi_*)L''_{xx}(x_*, u_*)y = -\sigma(I_n - \Pi_*)y. \tag{2.108}$$

If $\Pi_* y \neq 0$ then $\sigma = -1$, by virtue of (2.108). If $\Pi_* y = 0$, then $y = (I_n - \Pi_*)y$ and the second equation of (2.108) may be rewritten in the form

$$(I_n - \Pi_*)L''_{xx}(x_*, u_*)(I_n - \Pi_*)y = -\sigma y,$$

i.e. σ is an eigenvalue of the matrix defined in the statement of the lemma. This matrix is symmetric since Π_* and L''_{xx} are symmetric matrices. Moreover, this matrix is positive semi-definite. Indeed, for any w

$$(w, (I_n - \Pi_*)L''_{xx}(I_n - \Pi_*)w) = (z, L''_{xx} z)$$

where

$$z = (I_n - \Pi_*)w, \quad L''_{xx} = L''_{xx}(x_*, u_*).$$

But $f'(x_*)z = f'(x_*)(I_n - \Pi_*)w = 0$ because of equation (2.104); thus

$$(z, L''_{xx}z) \geq 0$$

because of the third of the main assumptions, and equality to zero is only possible if $z = (I_n - \Pi_*)w = 0$. From the symmetry of the given matrix, it follows that its eigenvalues and eigenvectors are real. Since $y \neq 0$, $\Pi_* y = 0$ implies that $(I_n - \Pi_*)y \neq 0$ and thus it follows from equation (2.108) that

$$-\sigma(y, y) = (y, (I_n - \Pi_*)L''_{xx}(I_n - \Pi_*)y) = (y, L''_{xx}y) > 0.$$

Thus, $\sigma \neq 0$ and so $-\sigma > 0$ and is a strictly positive eigenvalue of the given matrix.

It remains to establish the number of eigenvalues of the matrix $p'(x_*)$ equal to -1. By the construction of the matrix Π_* (see (2.104))

$$\Pi_* f_i'^*(x_*) = f_i'^*(x_*), \quad i \in I_*$$

i.e. the matrix $(I_n - \Pi_*)$ has $|I_*|$ zero eigenvalues. Consequently, the matrix $(I_n - \Pi_*)L''_{xx}(I_n - \Pi_*)$ has $|I_*|$ zero eigenvalues. On the other hand, as we have seen, the matrix $p'(x_*)$ has n nonzero eigenvalues. Thus, exactly $|I_*|$ of the eigenvalues of $p'(x_*)$ are equal to -1. This completes the proof of the lemma. □

This characteristic of the eigenvalues enables us to describe local characteristics of the rate of convergence of the linearization method using a theorem due to Ostrovskij [12, p.130].

Suppose that $g(x)$ is a mapping from $\mathbb{R}^n$ to $\mathbb{R}^n$ which is continuously differentiable and has fixed point x_*, $x_* = g(x_*)$. If all the eigenvalues of the matrix $g'(x_*)$ have modulus less than unity then there exists a neighbourhood of the point x_* such that the iteration process $x_{k+1} = g(x_k)$ converges to the point x_* and for k sufficiently large and any $\epsilon > 0$, the bound

$$\|x_k - x_*\| \leq C(q + \epsilon)^k$$

holds, where C depends only on ϵ and $q = \max_j |\lambda_j|$ where the λ_j are the eigenvalues of the matrix $q'(x_*)$.

Theorem 2.14 *Suppose that the main assumptions 1–3 of Section 2.3.1 are satisfied and that $p(x)$ is a solution of the auxiliary problem (2.91). Then there exist a sufficiently small neighbourhood of the point x_* (solution of problem (2.86)) and a sufficiently small $\alpha > 0$ such that the iterative process*

$$x_{k+1} = x_k + \alpha p(x_k) \tag{2.109}$$

converges to x_* from this neighbourhood and the bound

$$\|x_k - x_*\| \leq C(q+\epsilon)^k, \quad q < 1$$

holds.

Proof. If we set $g(x) = x + \alpha p(x)$, then x_* is a fixed point of the mapping g since $p(x_*) = 0$. Here

$$g'(x_*) = I_n + \alpha p'(x_*)$$

Thus, according to Lemma 2.4, the eigenvalues of the matrix $g'(x_*)$ have the form $1-\alpha$ or $1-\alpha\sigma$, where $\sigma > 0$ are nonzero eigenvalues of the matrix $A = (I_n - \Pi_*)L''_{xx}(I_n - \Pi_*)$. If $\alpha > 0$ is chosen so that $1-\alpha > -1$ and $1-\alpha\sigma > -1$, i.e. $\alpha < \min\{2, 2/\sigma_{\max}\}$, where $\sigma_{\max}$ is the maximum eigenvalue of the matrix A, then all the values $1-\alpha$ and $1-\alpha\sigma$ will lie within the interval $[-1,+1]$, whence by Ostrovskij's theorem, the iterative process (2.109) will converge locally. □

Generally speaking, this theorem does not give an accurate bound for the linearization method, since the process (2.109) has a constant step α, which is chosen using a method which we have not yet mentioned. However, Theorem 2.14 may serve as a guide and gives a general idea of the nature of the convergence. In particular, it has the following curious and important corollary.

Corollary *If $|I_*| = n$, i.e. if n linearly independent constraints are active in the solution, then for $\alpha = 1$ the rate of convergence of the process is superlinear.*

Proof. If $|I_*| = n$ and the vectors $f_i'(x_*)$ are linearly independent, then the matrix $f'(x_*)$ is invertible, whence

$$\begin{aligned} \Pi_* &= f'^*(x_*)[f'(x_*)f'^*(x_*)]^{-1}f'(x_*) \\ &= f'^*(x_*)(f'^*(x_*))^{-1}(f'(x_*))^{-1}f'(x_*) \\ &= I_n \end{aligned}$$

Thus, according to formula (2.107)

$$p'(x_*) = -\Pi_* = -I_n$$

so that all the eigenvalues of this matrix are equal to -1. Thus, $1-\alpha = 0$ with $\alpha = 1$, i.e.

$$g'(x_*) = I_n - p'(x_*) = 0$$

and we have a bound for the convergence of the process (2.107) in the form

$$\|x_k - x_*\| \leq C(\epsilon)\epsilon^k$$

showing that x_k converges to x_* more rapidly than any geometric progression.

2.3.3 Preliminary Lemmas

In this section we shall formulate an algorithm for solving problem (2.86) which has a high rate of convergence. Here, we shall prove a number of auxiliary results, which on the one hand form the basis for the formulation of the algorithm and on the other hand enable us to provide a rigorous justification of the latter.

We let $A(x, h)$ denote the $n \times n$ matrix with elements

$$h^{-2}\left[L(x+(e_i+e_j)h, u(x)) - L(x+e_ih, u(x)) - L(x+e_jh, u(x)) + L(x, u(x))\right]$$

where e_i are the unit vectors of the space $\mathbb{R}^n$. It is easy to see that $A(x, h)$ is an approximation to the matrix $L''_{xx}(x, u(x))$ with accuracy of the order h. We note that throughout this section, we shall consider a sufficiently small neighbourhood of the point x_*, whence, according to Lemma 2.3, the vector $u(x)$ will be uniquely defined. The same may be said about all the other values related to the solution of the auxiliary problem (2.91) which occur in what follows.

Suppose that $B(x, h) = -[\Pi(x) + (I - \Pi(x))A(x, h)]$.

Lemma 2.5 *In some neighbourhood of the point x_* the bound*

$$\|p'(x_*) - B(x, h)\| \leq C(\|x - x_*\| + h)$$

holds.

Proof. In fact, $A(x, h)$ differs from $L''_{xx}(x, u(x))$ by a value of the order h and the expression

$$-[\Pi(x) + (I - \Pi(x))L''_{xx}(x, u(x))]$$

is differentiable with respect to x and thus, by virtue of formula (2.107) differs from $p'(x_*)$ by a value of the order $\|x - x_*\|$. □

Lemma 2.6 *In some neighbourhood of the point x_*, the bound*

$$\|(p'(x_*))^{-1}p(x) - B^{-1}(x, h)p(x)\| \leq C(\|x - x_*\| + h)\|x - x_*\|$$

holds.

Proof. Suppose that

$$p'(x_*)y = -p(x), \quad B(x, h)\overline{y} = -p(x).$$

Then it is easy to show that

$$\overline{y} - y = (p'(x_*))^{-1}(B(x, h) - p'(x_*))B^{-1}(x, h)p(x). \tag{2.110}$$

Since $p(x_*) = 0$ and the vector $p(x)$ is differentiable, $\|p(x)\|$ is of the same order of magnitude as $\|x - x_*\|$. The assertion of the lemma now follows from this fact, Lemma 2.5 and (2.110). □

Lemma 2.7 *In a sufficiently small neighbourhood of the point x_*, the bound*

$$\|x - B^{-1}(x, h)p(x) - x_*\| \leq C(2\|x - x_*\| + h)\|x - x_*\|$$

holds.

Proof. Since the derivative of the vector $p(x)$ satisfies a Lipschitz condition, it is easy to obtain the bound

$$\|p(x) - p'(x_*)(x - x_*)\| \le C\|x - x_*\|^2. \tag{2.111}$$

To simplify the calculations, without loss of generality, in what follows, we shall assume that $x_* = 0$. Then

$$\begin{aligned}\|x - B^{-1}(x,h)p(x)\| &\le \|x - (p'(x_*))^{-1}p(x)\| \\ &\quad + \|(p'(x_*))^{-1}p(x) - B^{-1}(x,h)p(x)\| \\ &\le \|x - (p'(x_*))^{-1}[p'(x_*)x + (p(x) - p'(x_*)x)]\| \\ &\quad + C(\|x\| + h)\|x\| \\ &\le C\|x\|^2 + C(\|x\| + h)\|x\|\end{aligned}$$

where we have used Lemma 2.6 and the bound (2.111). □

Lemma 2.8 *Suppose that y is a solution of the system of equations*

$$B(x,h)y = -p(x). \tag{2.112}$$

Then y is a solution of the system of equations

$$\begin{aligned} A(x,h)y + (f_0'(x))^* + (f'(x))^*v &= 0 \\ f'(x)y + f(x) &= 0 \end{aligned} \tag{2.113}$$

in the unknowns $y \in \mathbb{R}^n$ and $v \in \mathbb{R}^l$, where $l = |I_*|$ is the number of elements of the set I_*. The converse is also true.

Proof. From the expression for $B(x,h)$ and the formula (2.98) for $p(x)$, it follows that equation (2.112) may be rewritten in the form

$$\Pi(x)y + (I - \Pi(x))A(x,h)y = -(I - \Pi(x))(f_0'(x))^* - f'^*(x)v(x). \tag{2.114}$$

It is easy to check that the following equations are satisfied

$$\Pi_*(x) = \Pi(x), \quad \Pi^2(x) = \Pi(x), \quad \Pi(x)(I - \Pi(x)) = 0, \quad \Pi(x)f'^*(x) = f'^*(x)$$

thus, multiplying (2.114) by $\Pi(x)$ and $(I - \Pi(x))$, we obtain

$$\Pi(x)y = -f'^*(x)v(x) \tag{2.115}$$

$$(I - \Pi(x))[A(x,h)y + (f_0'(x))^*] = 0. \tag{2.116}$$

Multiplying equation (2.115) by $f'(x)$ and by virtue of formulae (2.99) and (2.100), we obtain

$$f'(x)y = -f(x). \tag{2.117}$$

Moreover, the operator $\Pi(x)$ is the projection operator onto the subspace spanned by the vectors $(f_i'(x))^*$, $i \in I_*$. Equation (2.116) implies that the vector $A(x,h)y + (f_0'(x))^*$ lies in this subspace, i.e.

$$A(x,h)y + (f_0'(x))^* = -f'^*(x)v. \tag{2.118}$$

Equations (2.117) and (2.118) prove the lemma. □

2.3.4 The Linearization Algorithm and Acceleration of Convergence

We shall now show how the linearization method may be modified to obtain a high rate of convergence. We shall assume that the conditions for the convergence of the linearization method given in Section 2.1 are satisfied together with the main assumptions 1–3 of Section 2.3.1.

Suppose that $N > 0$, $\delta > 0$ an initial approximation x_0 and numbers $0 < \gamma < 1$, $0 < \epsilon < 1$, $h > 0$ have been chosen. We set $C_0 = +\infty$.

We now describe a general step of the algorithm. Suppose that x_k and C_k have already been constructed.

1. By solving problem (2.91) for $x = x_k$, we compute $p_k = p(x_k)$ and $u_k^i = u^i(x_k)$.

2. If $\|p_k\| \leq C_k$, then we set $h_k = \min(h, \|p_k\|)$ and compute y_k by solving the system of equations

$$A(x_k, h_k)y + (f_0'(x_k))^* + \sum_{i \in I(x_k)} v^i (f_i'(x_k))^* = 0$$
$$f_i'(x_k)y + f_i(x_k) = 0, \quad i \in I_\delta^0(x_k). \tag{2.119}$$

   If the system does not have a solution, then we set $C_{k+1} = \gamma \|p_k\|$ and move to stage 4.

   If $\|p_k\| > C_k$, then we set $C_{k+1} = C_k$ and move to stage 4.

3. If the system (2.119) has a solution y_k, then we set

$$\overline{x} = x_k + y_k \tag{2.120}$$

   and compute the vector $p(\overline{x})$ by solving problem (2.91) for $x = \overline{x}$.

   If

$$\|p(\overline{x})\| \leq \gamma \|p_k\| \tag{2.121}$$

   then we set $x_{k+1} = \overline{x}$ and $C_{k+1} = \gamma \|p_k\|$ and move to stage 1.

   If $\|p(\overline{x})\| > \gamma \|p_k\|$ then we set $C_{k+1} = \gamma \|p_k\|$ and move to the next stage.

4. We initially set $\alpha = 1$ and divide it by two until the inequality

$$f_0(x_k + \alpha p_k) + NF(x_k + \alpha p_k) \leq f_0(x_k) + NF(x_k) - \alpha\epsilon \|p_k\|^2 \tag{2.122}$$

   is satisfied. We set $x_{k+1} = x_k + \alpha p_k$. The algorithm now returns to stage 1.

Theorem 2.15 *Suppose that the main assumptions of Section 2.1.1 are satisfied. Then the above algorithm generates a sequence x_k for which $F(x_k) \to 0$ and the necessary conditions for an extremum are satisfied at any limit point. If, in addition, the main assumptions of Section 2.3.1 are satisfied and the solution x_* of problem (2.86) is the only point at which the necessary conditions* *for an extremum are satisfied, then the sequence* x_k *converges quadratically to* x_**, where, for* k *sufficiently large,* x_k *is updated to* x_{k+1} *according to stage 3 of the algorithm.*

Proof. From the construction of the algorithm, x_k is updated to x_{k+1} via formulae (2.119) and (2.120) or using the linearization method with step chosen according to formula (2.122). Here, the whole sequence $\{x_k\}$ is divided into segments $\{x_{k_j}, x_{k_{j+1}}, \ldots, x_{q_j}, \ldots x_{k_{j+1}-1}\}$, $j = 1, \ldots, q_j \geq k_j$ with the following properties. For the indices $k = k_j, k_{j+1}, \ldots q_j - 1$, x_k is updated to x_{k+1} via formulae (2.119) and (2.120), where

$$\|p_{k+1}\| \leq \gamma \|p_k\|, \quad k = k_j, \ldots, q_j - 1.$$

Thus,

$$\|p_{q_j}\| \leq \gamma^{(q_j - k_j)} \|p_{k_j}\|.$$

For the indices k from q_j to $k_{j+1} - 1$ the linearization method applies. According to the algorithm $C_{q_j} = \gamma \|p_{q_j}\|$ and $C_k = C_{q_j}$ for all $k \geq q_j$ until the inequality $\|p_{k_{j+1}}\| \leq C_{k_{j+1}} = C_{q_j} = \gamma \|p_{q_j}\|$ is satisfied. This inequality is first satisfied for the index k_{j+1}. Study of the linearization method has shown that this inequality is necessarily satisfied. Thus,

$$\|p_{k_{j+1}}\| \leq \gamma \|p_{q_j}\| \leq \gamma^{(q_j - k_j)+1} \|p_{k_j}\|.$$

Whence, it follows that $\|p_{k_j}\| \to 0$ and so, according to Section 2.1 any convergent subsequence of the sequence $\{x_{k_j}\}$ has a limit point at which the constraints and the necessary conditions for an extremum are satisfied. This proves the first assertion of the theorem.

We shall now prove the second assertion. Let us consider the block matrix

$$\begin{pmatrix} L''_{xx}(x_*, u_*) & f'^*(x_*) \\ f'(x_*) & 0 \end{pmatrix}. \tag{2.123}$$

We shall show that this matrix is nonsingular.

In fact, if this matrix is singular, then there would exist vectors $y \subset \mathbb{R}^n$, $v \in \mathbb{R}^l$, not all zero, satisfying the system

$$\begin{aligned} L''_{xx}(x_*, u_*)y + f'^*(x_*)v &= 0 \\ f'(x_*)y &= 0 \end{aligned}$$

Carrying out scalar multiplication of the first of these equations by y and taking into account the second equation, we obtain:

$$\begin{aligned} (y, L''_{xx}(x_*, u_*)y) &= 0 \\ f'(x_*)y &= 0; \end{aligned}$$

whence it follows that $y = 0$, since equations (2.89) and (2.90) are satisfied. But then, for v not equal to zero, we have the equation $f'^*(x_*)v = 0$, which contradicts the fact that the vectors $f'_i(x_*)$, $i \in I_*$ are linearly independent.

According to Lemma 2.3, $I_\delta^0(x) = I_*$ if x is sufficiently close to x_*. Thus, the matrix of the system (2.119) will have the form

$$\begin{pmatrix} A(x_k, h_k) & f'^*(x_k) \\ f'(x_k) & 0 \end{pmatrix}. \tag{2.124}$$

provided x_k is close to x_*. Taking into account the fact that $h_k = \min(h, \|p_k\|) \le \|p(x_k)\|$, whence h_k is also small, we deduce that the matrix of (2.124) is close to the matrix of (2.123) and is also nonsingular. Thus, the system (2.119) is solvable for sufficiently large $k = k_j$. In this case, according to Lemma 2.8, the system (2.119) is equivalent to the equation

$$B(x_k, h_k)y_k = -p(x_k)$$

so that formula (2.120) gives

$$\overline{x} = x_k - B^{-1}(x_k, h_k)p(x_k). \tag{2.125}$$

It now follows from Lemma 2.7 that

$$\|\overline{x} - x_*\| \le C(2\|x_k - x_*\| + h_k)\|x_k - x_*\|. \tag{2.126}$$

We note that, according to Lemma 2.3, in a neighbourhood of x_*,

$$p(x) = p'(x_*)(x - x_*) + \omega(x), \quad \|\omega(x)\| \le C\|x - x_*\|^2. \tag{2.127}$$

It can be shown that if the matrix of (2.123) is nonsingular then the matrix $p'(x_*)$ is also nonsingular. Thus, there exists a number $m > 0$ such that

$$\|p'(x_*)(x - x_*)\| \ge m\|x - x_*\|.$$

Whence

$$\begin{aligned} \|p(x)\| &\ge \|p'(x_*)(x - x_*)\| - \|\omega(x)\| \\ &\ge m\|x - x_*\| - C\|x - x_*\|^2 \\ &\ge (m - C\|x - x_*\|)\|x - x_*\|. \end{aligned}$$

Thus, for $\|x - x_*\|$ sufficiently small

$$\|p(x)\| \ge (m/2)\|x - x_*\|. \tag{2.128}$$

Let us now return to formula (2.126). Since $h_k \le \|p(x_k)\| \le C\|x_k - x_*\|$, by renaming the constants, formula (2.126) may be rewritten in the form

$$\|\overline{x} - x_*\| \le C\|x_k - x_*\|^2, \quad k = k_j. \tag{2.129}$$

Then

$$\begin{aligned} \|p(\overline{x})\| &\le C\|\overline{x} - x_*\| \\ &\le C_1\|x_k - x_*\|^2 \\ &\le ((2C_1/m)\|x_k - x_*\|)(m/2)\|x_k - x_*\| \\ &\le ((2C_1/m)\|x_k - x_*\|)p(x_k). \end{aligned}$$

If

$$(2C_1/m)\|x_k - x_*\| \le \gamma,$$

then, we finally obtain that, in a small neighbourhood of the point x_*

$$\|p(x_{k+1})\| = \|p(\overline{x})\| \le \gamma\|p(x_k)\|.$$

Thus, if $k = k_j$ is sufficiently large then the algorithm provides for updating from x_k to x_{k+1} by formulae (2.119) and (2.120) and the bound (2.129) holds. This bound also shows that if x_k lies in the required neighbourhood of x_* and $C\|x_k - x_*\| < 1$ then x_{k+1} will also lie in this neighbourhood.

Let us summarize the above. For j sufficiently large the point x_{k_j} lies in a neighbourhood of the point x_* in which all the assumptions under which the previous calculations were carried out are valid. The equation

$$\|x_{k+1} - x_*\| \le C\|x_k - x_*\|^2 \tag{2.130}$$

holds and the point x_{k+1} again lies in the required neighbourhood. In fact, we have proved that, from some k on, formula (2.130) holds, showing that the process converges quadratically. □

2.3.5 Linear Transformations of the Problem

We shall now make a small digression and consider how a number of parameters in problem (2.86) and in the auxiliary problem (2.91) vary under coordinate transformations and multiplication of the functions f_i by positive constants.

We recall that if x_* is a solution of problem (2.86) and u_* is the corresponding Lagrange multiplier vector, then

$$\begin{aligned} f_0'(x_*) + \sum_{i=1}^{m} u_*^i f_i'(x_*) &= 0 \\ u_*^i \ge 0, \;\; u_*^i f_i(x_*) &= 0, \;\; i \in I. \end{aligned} \tag{2.131}$$

Suppose that $\tilde{f}_i = a_i f_i$, $a_i > 0$, $i = 0, \ldots, m$. Clearly, x_* is a solution of the problem

$$\min_x \{\tilde{f}_0(x) | \tilde{f}_i(x) \le 0, \; i \in I\}.$$

On the other hand, $\tilde{f}_i' = a_i f_i'$ and (2.131) may be easily transformed to the form

$$\begin{aligned} \tilde{f}_0'(x_*) + \sum_{i=1}^{m} \tilde{u}_*^i \tilde{f}_i'(x_*) &= 0 \\ \tilde{u}_*^i \geq 0,\ \ \tilde{u}_*^i \tilde{f}_i(x_*) &= 0,\ \ i \in I \end{aligned}$$

where

$$\tilde{u}_*^i = (a_0/a_i)u_i. \tag{2.132}$$

Thus, multiplying all the functions f_i by positive constants leads to the transformation (2.132) of the Lagrange multipliers.

Next we consider what happens if we carry out a coordinate transformation

$$x = Ay \tag{2.133}$$

where A is a nonsingular matrix. We recall that if $f(x)$ is a twice differentiable function with gradient $f'(x)$ (row vector) and matrix of second derivatives $f''(x)$ then for the function $\tilde{f}(y) = f(Ay)$ we have

$$\tilde{f}'(y) = f'(x)A \tag{2.134}$$

$$\tilde{f}''(y) = A^* f''(x)A \tag{2.135}$$

where the dashes denote the derivatives with respect to y.

Let us now consider the problem

$$\min_y \{\tilde{f}_0(y) | \tilde{f}_i(y) \leq 0,\ i \in I\}. \tag{2.136}$$

Clearly, this has solution $y_* = A^{-1}x_*$. If we multiply (2.131) on the right by A, then, taking into account (2.134), we immediately obtain the necessary conditions for an extremum for problem (2.136). Thus, in the case of coordinate transformations the Lagrange multipliers are unchanged.

Let us write down the auxiliary problem of the linearization method for problem (2.136). It is easy to see that if x and y are related by equation (2.133) then $I_\delta(y) = I_\delta(x)$, and, by virtue of (2.134), the auxiliary problem has the form

$$\min_{\tilde{p}} \{f_0'(x)A\tilde{p} + 1/2(\tilde{p}, \tilde{p}) | f_i'(x)A\tilde{p} + f_i(x) \leq 0,\ i \in I_\delta(x)\}. \tag{2.137}$$

In deriving (2.137), we used the fact that $\tilde{f}'$ is a row vector, whence

$$(\tilde{f}'(y), \tilde{p}) = f'(x)A\tilde{p},$$

where the product on the right-hand side is calculated by matrix multiplication. We denote the solution of problem (2.137) by $\tilde{p}(y)$. In (2.137) we set $p = A\tilde{p}$. Then (2.137) takes the form

$$\min_p \{f_0'(x)p + 1/2(A^{-1}p, A^{-1}p) | f_i'(x)p + f_i(x) \leq 0,\ i \in I_\delta(x)\}. \tag{2.138}$$

Setting

$$C = (A^{-1})^* A^{-1} \tag{2.139}$$

we obtain the problem

$$\min_p\{f_0'(x)p + 1/2(p, Cp) | f_i'(x)p + f_i(x) \le 0,\ i \in I_\delta(x)\}. \tag{2.140}$$

Therefore, if $p_C(x)$ is a solution of problem (2.140) there is a one-to-one connection

$$p_C(x) = A\tilde{p}(y). \tag{2.141}$$

We also note that, by virtue of the above, the Lagrange multipliers for problems (2.137) and (2.140) are the same.

Let us now suppose that we have carried out the transformation (2.133) and apply the linearization method to problem (2.136). Then, the point y is updated to a new point $\overline{y} = y + \alpha\tilde{p}$ by choosing the step α by dividing the initial value $\alpha = 1$ by two until the inequality

$$\tilde{f}_0(y + \alpha\tilde{p}(y)) + N\tilde{F}(y + \alpha\tilde{p}(y)) \le \tilde{f}_0(y) + N\tilde{F}(y) - \alpha\epsilon\|\tilde{p}(y)\|^2$$

is satisfied. But $\tilde{f}_0(y) = f_0(Ay)$, $\tilde{F}(y) = F(Ay)$ and $x = Ay$ so that the last inequality takes the form

$$f_0(Ay + \alpha A\tilde{p}(y)) + NF(Ay + \alpha A\tilde{p}(y)) \le f_0(Ay) + NF(Ay) - \alpha\epsilon\|\tilde{p}(y)\|^2$$

or, by virtue of (2.139) and (2.141),

$$\begin{aligned} f_0(x + \alpha p_C(x)) + NF(x + \alpha p_C(x)) \ \le\ & f_0(x) + NF(x) \\ & -\alpha\epsilon(p_C(x), Cp_C(x)). \end{aligned} \tag{2.142}$$

Thus, in fact, there is no need to transfer to new coordinates. Instead, all the calculations may be carried out in the old coordinates, by calculating $p_C(x)$ as the solution of problem (2.140) and transferring to the new point $\overline{x} = x + \alpha p_C(x)$, choosing the step α from the condition (2.142).

On the other hand, if an arbitrary strictly positive matrix C is chosen (i.e. $(p, Cp) > 0$ for $p \ne 0$) then it is known [46] that it may be represented in the form $C = B^*B$ (here, B may also be chosen to be an upper triangular matrix). Setting $A = B^{-1}$, we see that the minimization process defined by formulae (2.140) and (2.142) is equivalent to the standard linearization method in the coordinates y which are related to x by the equations

$$x = Ay = B^{-1}y,\ \ C = B^*B. \tag{2.143}$$

It is interesting to compute the matrix $\tilde{p}'(y_*)$ in the new coordinates y. We use formula (2.107) for this. In the new coordinates, taking into account the fact that the Lagrange multipliers are unchanged, we have

$$L''_{yy}(y_*, u_*) = A^*L''_{xx}(x_*, u_*)A,$$

and $\tilde{\Pi}_*$ is the projection operator onto the subspace orthogonal to the subspace $\tilde{f}'(y_*)\tilde{p} = 0$ or $f'(x_*)A\tilde{p} = 0$. Thus, formula (2.107) takes the form

$$\tilde{p}'(y_*) = -[\widetilde{\Pi}_* + (I_n - \widetilde{\Pi}_*)A^* L''_{xx}(x_*, u_*)A]. \tag{2.144}$$

Let us now assume that the matrix $L''_{xx}(x_*, u_*)$ is strictly positive-definite. Then it may be represented in the form

$$L''_{xx}(x_*, u_*) = (A_0^{-1})^* A_0^{-1}. \tag{2.145}$$

If now, with a coordinate transformation, we set $A = A_0$, then

$$A_0^* L''_{xx}(x_*, u_*)A_0 = I_n$$

and formula (2.144) takes the form

$$\tilde{p}'(y_*) = -[\widetilde{\Pi}_* + (I_n - \widetilde{\Pi}_*)I_n] = -I_n.$$

Therefore, if the coordinate transformation $x = A_0 y$ is chosen according to condition (2.145) then $\tilde{p}'(y_*) = -I_n$.

2.3.6 Modifications of the Linearization Method

From Section 2.3.5 it follows that the auxiliary problem (2.91) is not necessarily used in implementations of the linearization method, which may be based on resolution of problem (2.140) and selection of the step according to formula (2.142). We shall go further and change the matrix C from iteration to iteration.

Modified Linearization Algorithm Suppose that x_0 is chosen, that C_k are strictly positive-definite symmetric matrices, $k = 0, 1, \dots, N$, and that $\delta > 0$ and $0 < \epsilon < 1$. If x_k has already been constructed, then x_{k+1} is constructed according to the following rule:

1. We solve the quadratic programming problem

   $$\min_p\{(f_0'(x_k), p) + 1/2(p, C_k p) | (f_i'(x_k), p) + f_i(x_k) \leq 0,\ i \in I_\delta(x_k)\} \tag{2.146}$$

   Suppose that it has a solution p_k with corresponding Lagrange multipliers u_k^i.

2. Beginning with $\alpha = 1$, we divide α by two until the following inequality is satisfied:

   $$f_0(x_k + \alpha p_k) + NF(x_k + \alpha p_k) \leq f_0(x_k) + NF(x_k) - \alpha\epsilon(p_k, C_k p_k). \tag{2.147}$$

   We obtain the value for the step $\alpha_k = \alpha$.

3. We set $x_{k+1} = x_k + \alpha_k p_k$ and return to stage 1.

Theorem 2.16 *Suppose that the following conditions are satisfied:*

a) The set

$$\Omega_0 = \{f_0(x) + NF(x) \leq f_0(x_0) + NF(x_0)\}$$

is compact.

b) The gradients $f_i'(x)$ in this region satisfy a Lipschitz condition with constant L.

c) There exist numbers $M \geq m > 0$ such that

$$m\|p\|^2 \leq (p, C_k p) \leq M\|p\|^2.$$

d) The inequality

$$\sum_i u_k^i \leq N$$

is satisfied in all the iterations.

Then $F(x_k) \to 0$ and any limit point of the sequence $\{x_k\}$ satisfies the constraints of problem (2.86) and the necessary conditions for an extremum.

Proof. The necessary and sufficient conditions for a minimum for a solution p_k of problem (2.146) have the form

$$\begin{aligned} C_k p_k^* + f_0'(x_k) + \sum_{i \in I_\delta(x_k)} u_k^i f_i'(x_k) &= 0 \\ u_k^i \geq 0, \;\; u_k^i[(f_i'(x_k), p_k) + f_i(x_k)] &= 0, \;\; i \in I_\delta(x_k). \end{aligned} \tag{2.148}$$

Carrying out scalar multiplication of the first of these equations by p_k, we obtain the equation

$$(f_0'(x_k), p_k) - \sum_{i \in I_\delta(x_k)} u_k^i f_i(x_k) - (p_k, C_k p_k) \tag{2.149}$$

Then, repeating all the calculations exactly as in Theorem 2.1, with the sole difference that equation (2.149) is used instead of equation (2.7), we obtain

$$\begin{aligned} f_0(x_k + \alpha p_k) + NF(x_k + \alpha p_k) &\leq f_0(x_k) + NF(x_k) - \alpha\epsilon(p_k, C_k p_k) \\ &\leq f_0(x_k) + NF(x_k) - \alpha\epsilon m\|p_k\|^2 \end{aligned} \tag{2.150}$$

for $0 \leq \alpha \leq \overline{\alpha}_k$,

$$\overline{\alpha}_k = \min\left(1, \frac{\delta}{F(x_k) + K\|p_k\|}, \frac{1-\epsilon}{(N+1)L}\right).$$

Equation (2.150) enables us to repeat all the arguments used in Theorem 2.1 to complete the proof of Theorem 2.16. □

The statement of Theorem 2.16 includes the condition d) which we found impossible to verify before. However, this may be efficiently verified during the process and the chosen initial value of N may be corrected as in the discussion of the main algorithm of the linearization method in Section 2.1.

We now use the results obtained to construct an algorithm with a high rate of convergence for the convex programming problem. The basic idea behind this is the following. According to the above, if the point x_* and the Lagrange multipliers were known, it would be possible to transfer to new coordinates y, taking a transformation matrix A_0 based on formula (2.145). Here, the matrix $\tilde{p}'(y_*)$ is equal to $-I_n$, i.e. all its eigenvalues are equal to -1, and thus, according to Section 2.3.2, the process $y_{k+1} = y_k + \tilde{p}(y_k)$ would converge to the point y_* more rapidly than any geometric progression from a neighbourhood of the point y_*. But the points x_* and u_* are unknown. Thus, we shall try to use $L''_{xx}(x_k, u_{k-1})$ at each step, instead of $L''_{xx}(x_*, u_*)$, where u_{k-1} is the Lagrange multiplier vector from the previous iteration. Here, global convergence will be guaranteed by the fact that we shall only use the modified algorithm studied above, while local superlinear convergence is guaranteed by the technique which has already been used to construct the algorithm in Section 2.3.4.

Algorithm for Convex Programming Problems Suppose that x_0, $N > 0$, $\delta > 0$, $0 < \epsilon < 1$ and $0 < \gamma < 1$ are given. We set $C_0 = +\infty$ and $u_{-1} = 0$, $u_{-1} \in \mathbb{R}^m$.

Suppose that the point x_k, the Lagrange multiplier vector u_{k-1} and the number C_k have already been constructed.

1. We solve the problem

$$\begin{aligned}\min_p\{&(f_0'(x_k), p) + 1/2(p, L''_{xx}(x_k, u_{k-1})p) :\\ &(f_i'(x_k), p) + f_i(x_k) \leq 0,\ i \in I_\delta(x_k)\}.\end{aligned} \tag{2.151}$$

   Suppose that this has solution p_k with corresponding Lagrange multipliers u_k^i, $i \in I_\delta(x_0)$. We set $u_k^i = 0$ for $i \notin I_\delta(x_k)$.

2. If $\|p_k\| > C_k$, then we set $C_{k+1} = C_k$ and move to stage 4. If $\|p_k\| \leq C_k$, we transfer to stage 3.

3. We solve problem (2.151) in which we replace x_k and u_{k-1} by $\overline{x} = x_k + p_k$ and u_k, respectively. If the solution $\overline{p}$, $\overline{u}$ is such that $\|\overline{p}\| \leq \gamma\|p_k\|$, then we set

$$x_{k+1} = x_k + p_k,\quad C_{k+1} = \gamma\|p_k\|$$

   and move to stage 1. (We note that now problem (2.151) for the new data has actually already been solved, since $p_{k+1} = \overline{p}$ and $u_{k+1} = \overline{u}$ and there is no need to solve it again.)

   But if $\|\overline{p}\| > \gamma\|p_k\|$, then we set $C_{k+1} = \gamma\|p_k\|$ and transfer to stage 4.

4. Beginning with $\alpha = 1$, we divide α by two until the following inequality is satisfied:

$$f_0(x_k + \alpha p_k) + NF(x_k + \alpha p_k) \\ \leq f_0(x_k) + NF(x_k) - \alpha\epsilon(p_k, L''_{xx}(x_k, u_{k-1})p_k);$$

thus, we calculate α_k. We set $x_{k+1} = x_k + \alpha_k p_k$. We return to stage 1.

Stopping criterion: $\|p_k\| \leq w$, where w is the given accuracy.

We now formulate conditions which guarantee convergence of the algorithm.

Theorem 2.17 *Suppose that the $f_i(x)$, $i = 0, 1, \dots, m$ are twice continuously differentiable convex functions and that there exists a number $m > 0$ such that*

$$(p, f_0''(x)p) \geq m\|p\|^2 \tag{2.152}$$

for all x and p. Suppose also that at the minimum x_*, the gradients of the active constraints are linearly independent and the corresponding Lagrange multipliers are positive and that the inequality

$$\sum_{i=1}^{m} u_k^i \leq N$$

is satisfied for the sequence $\{x_k\}$ generated by the algorithm. Then the sequence $\{x_k\}$ converges to the point x_* more rapidly than any geometric progression.

Proof. The assumptions of the theorem enable us to draw a number of immediate conclusions. Because of the condition of strong convexity (2.152) of the function $f_0(x)$, this problem has a unique minimum x_*. In addition, this condition also implies that $f_0(x) \to \infty$ as $x \to \infty$, whence that the set $\{x | f_0(x) + NF(x) \leq C\}$ is compact for any N and C, since $F(x) \geq 0$. Moreover, because the functions $f_i(x)$, $i = 1, \dots, m$ are convex

$$(p, f_i''(x)p) \geq 0.$$

By virtue of (2.152), we obtain

$$(p, L''_{xx}(x_k, u_{k-1})p) \geq (p, f_0''(x)p) \geq m\|p\|^2.$$

Since the Lagrange multipliers u_k are uniformly bounded the inequality

$$(p, L''_{xx}(x_k, u_{k-1})p) \leq M\|p\|^2$$

holds.

Thus, all the conditions of Theorem 2.16 are satisfied, if in it we set $C_k = L''_{xx}(x_k, u_{k-1})$. Thus, if we were to apply the modified linearization algorithm it would generate a sequence converging to x_*.

We now note that the conditions of the theorem guarantee that the main assumptions 1–3 of Section 2.3.1 are satisfied so that Lemmas 2.2 and 2.3 hold and we may use the results of these lemmas in what follows.

Let us consider the proof of convergence. We show that the sequence C_k converges to zero. In fact, C_k decreases by a factor of at least γ whenever $\|p_k\| \leq C_k$. If the numbers C_k do not converge to zero, this means that they are constant from some instant on and that always $\|p_k\| > C_k$. But then the algorithm uses only stages 1 and 4, i.e. it behaves like the modified linearization method, which converges, as we saw above. But then, contrary to the assumptions, $p_k \to 0$ and thus at some instant $\|p_k\| \leq C_k$, after which $C_{k+1} = \gamma\|p_k\| \leq \gamma C_k$, which contradicts the fact C_k is constant.

Thus, $C_k \to 0$. Let J be the set of indices k for which C_k is changed. For these indices $\|p_k\| \leq C_k$ and so p_k tends to zero as $k \to \infty$, $k \in J$.

Turning now to the proof of Theorem 2.1, which is actually the basis for the proof of Theorem 2.16, using the linear independence of the gradients of the active constraints, we obtain that

$$x_k \to x_*, \quad u_k \to u_* \tag{2.153}$$

as $k \to \infty$, $k \in J$.

We now show that, from some instant on, x_k will be updated to x_{k+1} via the formula

$$x_{k+1} = x_k + p_k \tag{2.154}$$

according to stage 3 of the algorithm.

We begin with the following remark: $k \in J$, if $\|p_k\| \leq C_k$ and in this case an attempt is made to take a step according to formula (2.154). If this is unsuccessful according to stage 3 of the algorithm the modified linearization method is used. If the total number of successful attempts to use formula (2.154) was finite then the modified linearization method was used all the time and $x_k \to x_*$ and $u_k \to u_*$ as $k \to \infty$. Thus, we must assume that an infinite number of attempts to use formula (2.154) were successful.

We denote the set of corresponding indices k by J^0. Clearly $J^0 \subseteq J$. But successful application of formula (2.154) according to stage 3 of the algorithm implies that

$$\|p_{k+1}\| \leq \gamma\|p_k\|.$$

Thus, $\|p_{k+1}\| \to 0$ as $k \to \infty$, $k \in J^0$. Whence, as before, we may conclude that

$$x_{k+1} \to x_*, \quad u_{k+1} \to u_* \tag{2.155}$$

as $k \to \infty$, $k \in J^0$.

Combining (2.153) and (2.155), we obtain

$$x_{k+1} \to x_*, \quad u_k \to u_*$$

as $k \to \infty$, $k \in J^0$; thus

$$L''_{xx}(x_{k+1}, u_k) \to L''_{xx}(x_*, u_*), \quad k \to \infty, \quad k \in J^0. \tag{2.156}$$

Let us now suppose that A_0 is calculated from formula (2.145). We carry out the coordinate transformation $x = A_0 y$.

Suppose that $\tilde{p}_0(y)$ is the solution of problem (2.137) for $A = A_0$. According to Section 2.3.4

$$\tilde{p}_0'(y_*) = -I_n.$$

On the other hand, if $p_0(x) = p_{C_0}(x)$ (the solution of (2.140) for $C = C_0 = L''_{xx}(x_*, u_*)$) then, according to formula (2.141)

$$p_0(x) = A_0 \tilde{p}_0(y) = A_0 \tilde{p}_0(A_0^{-1} x)$$

whence

$$p_0'(x_*) = A_0 \tilde{p}_0'(y_*) A_0^{-1} = -I_n. \tag{2.157}$$

Thus, it follows that

$$\begin{aligned} p_0(x) &= -I_n(x - x_*) + \omega(x - x_*) \\ &= -(x - x_*) + \omega(x - x_*) \end{aligned} \tag{2.158}$$

$$\lim_{x \to x_*} \omega(x - x_*)/\|x - x_*\| = 0. \tag{2.159}$$

Moreover, according to Lemmas 2.2 and 2.3, the set of active indices in the auxiliary problems (2.140) for $C = C_0$ and (2.151) for $k - 1 \in J^0$ and k sufficiently large (here we use (2.155)) is the same as I_*, the set of active indices in the main problem (2.86). Thus, according to the necessary conditions for an extremum (2.148), the following equations hold.

$$\begin{aligned} C_0 p_0(x_k) + f_0'^*(x_k) + f'^*(x_k) u_0(x_k) &= 0 \\ f'(x_k) p_0(x_k) + f(x_k) &= 0 \\ C_k p_k + f_0'^*(x_k) + f'^*(x_k) u_k &= 0 \\ f'(x_k) p_k + f(x_k) &= 0 \\ C_0 = L''_{xx}(x_*, u_*), \quad C_k &= L''_{xx}(x_k, u_{k-1}), \quad k - 1 \in J^0. \end{aligned}$$

Subtracting these equations we obtain

$$\begin{aligned} C_k(p_k - p_0(x_k)) + f'^*(x_k)(u_k - u_0(x_k)) - (C_0 - C_k) p_0(x_k) &= 0 \\ f'(x_k)(p_k - p_0(x_k)) &= 0. \end{aligned}$$

Using the fact that, considered as a system in $(p_k - p_0(x_k)), (u_k - u_0(x_k))$, this system is nonsingular, it is easy to obtain the bounds

$$\begin{aligned} \|p_k - p_0(x_k)\| &\le K \| C_0 - C_k \| \|p_0(x_k)\| \\ \|u_k - u_0(x_k)\| &\le K \| C_0 - C_k \| \|p_0(x_k)\|. \end{aligned} \tag{2.160}$$

Using these bounds, we obtain

$$\begin{aligned}
\overline{x} &= x_k + p_k = x_k + p_0(x_k) + (p_k - p_0(x_k)) \\
&= x_* + \omega(x_k - x_*) + (p_k - p_0(x_k)) \\
\|\overline{x} - x_*\| &\leq \|\omega(x_k - x_*)\| + \|p_k - p_0(x_k)\| \\
&\leq K\left[\frac{\|\omega(x_k - x_*)\|}{\|x_k - x_*\|} + \|C_0 - C_k\|\right]\|p_0(x_k)\| \\
&\leq K_1\left[\frac{\|\omega(x_k - x_*)\|}{\|x_k - x_*\|} + \|C_0 - C_k\|\right]\|x_k - x_*\| \qquad (2.161)
\end{aligned}$$

where we have used the fact that, by virtue of (2.158) and (2.159), the norms of $p_0(x)$ and $x - x_*$ have the same order of magnitude.

We shall not carry out further rather simple but tedious formal calculations. The important thing is that, for k sufficiently large, by virtue of formulae (2.156) and (2.159), the coefficient

$$q_k = K_1\left[\frac{\|\omega(x_k - x_*)\|}{\|x_k - x_*\|} + \|C_0 - C_k\|\right] \qquad (2.162)$$

is arbitrarily small and so $x_{k+1} = \overline{x} = x_k + p_k$ is a far better approximation to x_* than x_k. The closeness of p_k and $p_0(x_k)$ provided by formulae (2.160) guarantees that $\|p_{k+1}\| \leq \gamma\|p_k\|$. Thus, for k sufficiently large, the update from iteration to iteration is carried out via formula (2.154); whence, if $k-1 \in J^0$, then $k \in J^0$. Therefore, it follows that from some instant onwards, J^0 contains all the indices.

By virtue of (2.161),

$$\|x_{k+1} - x_*\| \leq q_k\|x_k - x_*\|$$

and formula (2.162) shows that $q_k \to 0$. This is superlinear convergence and the proof of the theorem is complete. □

As far as the conditions of Theorem 2.17, and those of Theorem 2.16 are concerned the impossibility of checking the boundedness of the Lagrange multipliers at an earlier stage is a nuisance. As previously mentioned, this disadvantage is to some extent compensated by the fact that this condition may be monitored during the process. The following theorem, which is easily proved from the formulae given in the proof of Theorem 2.17, may be useful.

Theorem 2.18 *Suppose that all the conditions of Theorem 2.17 are satisfied, except the condition about the boundedness of the Lagrange multipliers. Then the process constructed according to the previous algorithm converges to the solution x_* of problem (2.86) more rapidly than any geometric progression if the initial approximation x_0 is sufficiently close to x_*.*

In fact, if the initial approximation is sufficiently close to x_*, then formula (2.161) will hold and the algorithm will work using only formula (2.154), and the value of the constant N is not at all necessary.

There is yet another case in which neither the value of N nor the Lagrange multipliers affect the operation of the algorithm. That is, if the constraints $f_i(x)$, $i = 1, \ldots, m$ are linear in x, then

$$L''_{xx}(x,u) = f''_0(x).$$

Moreover, if $\delta = +\infty$ then all the constraints are involved in the auxiliary problem and, because of their linearity,

$$f_i(x+p) = (f'_i(x),p) + f_i(x) \le 0$$

so that if the initial point x satisfies these constraints, they will also be satisfied at all points $x + \alpha p$. Thus, $F(x+\alpha p) = 0$ and formula (2.150), which is used to choose the step, takes the form

$$f_0(x_k + \alpha p_k) \le f_0(x_k) - \epsilon\alpha(p_k, f''_0(x_k)p_k). \tag{2.163}$$

Thus, based on the above, we deduce the following theorem.

Theorem 2.19 *Suppose that $f_0(x)$ is a convex, twice continuously differentiable function and that $f_i(x)$, $i = 1,\dots,m$ are linear functions, i.e.*

$$f_i(x) = (a_i, x) - b_i.$$

Suppose also that

$$(p, f''_0(x), p) \ge m\|p\|^2, \quad m > 0.$$

Then the previous algorithm for convex programming, in which the problem

$$\min_p\{(f'_0(x_k),p) + 1/2(p, f''_0(x_k),p) | (a_i,p) + f_i(x) \le 0, \ i \in I\}$$

is solved instead of problem (2.151) and formula (2.163) is used to choose the step, converges more rapidly than any geometric progression from any initial approximation x_0 satisfying the constraints of the problem.

3. The Discrete Minimax Problem and Algorithms

3.1 The Discrete Minimax Problem

The discrete minimax problem, which is often met in practice, involves minimizing the maximum of a finite number of functions. Formally, if $I = \{1, \ldots, m\}$ is a finite set of indices and for each $i \in I$ a function $f_i(x)$ is given, then the problem consists of minimizing the function

$$F(x) = \max_{1 \le i \le m} f_i(x) \tag{3.1}$$

as x varies throughout the space $\mathbb{R}^n$.

A problem could be posed in which the argument was required to vary in some given region M. But this could deprive the problem of a number of specific features and could actually lead to the solution of a general nonlinear programming problem. The problem

$$\min_x \{F(x) | x \in M\}$$

is clearly equivalent to the problem

$$\min_{x,\beta} \{\beta | f_i(x) \le \beta,\ i \in I,\ x \in M\} \tag{3.2}$$

which in turn is equivalent to a general nonlinear programming problem, since by evident reasoning, any problem of the form (2.2) may be brought to the form (3.2) by the introduction of an additional variable.

Thus, the objective of this section is to construct algorithms to solve the problem,

$$\min_x \{F(x) | x \in \mathbb{R}^n\}. \tag{3.3}$$

As previously mentioned, this problem is equivalent to the problem

$$\min_{x,\beta} \{\beta | f_i(x) \le \beta,\ x \in \mathbb{R}^n\}. \tag{3.4}$$

We introduce certain notation and facts which will be needed in what follows.

Thus, in this section, we shall assume that the functions $f_i(x)$, $i \in I$ are continuously differentiable and that their gradients satisfy a Lipschitz condition on any compact set. In addition, we shall assume that the sets

$$C_\alpha = \{x | F(x) \leq \alpha\} \tag{3.5}$$

are compact.

Clearly, these assumptions guarantee the existence of a minimum of (3.3). Suppose that x_* is any such minimum. Then, according to Section 1.2.7 (Theorem 1.13) the following condition is satisfied at the point x_*: there exists a vector $u_* \in \mathbb{R}^m$ such that

$$\begin{aligned} \sum_{i=1}^m u_*^i f_i'(x_*) &= 0 \\ u_*^i \geq 0, \;\; u_*^i(f_i(x_*) - F(x_*)) &= 0, \;\; i = 1, \ldots, m \\ \sum_{i=1}^m u_*^i &= 1. \end{aligned} \tag{3.6}$$

We set

$$I_\delta(x) = \{i \in I | f_i(x) \geq F(x) - \delta\}, \;\; \delta > 0.$$

The main idea of what follows is the same as that behind the construction of the algorithms of Chapter 2. The difference is that we shall use the specification of problem (3.3), which enables us to dispense with a number of requirements on the original problem and algorithm which are difficult to verify.

3.1.1 The Auxiliary Problem

Suppose that x is a point and that A is a symmetric, strictly positive-definite matrix such that

$$(p, Ap) \geq m\|p\|^2. \tag{3.7}$$

Let us consider the auxiliary problem

$$\min_{p,\beta}\{\beta + 1/2(p, Ap) | (f_i'(x), p) + f_i(x) - \beta \leq 0, \; i \in I_\delta(x)\}. \tag{3.8}$$

Since the values $p = 0$ and $\beta \geq F(x)$ clearly satisfy the constraints of problem (3.8), this is always solvable.

By virtue of the constraints (3.8)

$$\beta \geq f_i(x) + (f_i'(x), p) \geq f_i(x) - \|p\|\|f_i'(x)\|, \;\; i \in I_\delta(x),$$

i.e.

$$\begin{aligned} \beta &\geq c_1(x) - \|p\| c_2(x) \\ c_1(x) &= \min_i\{f_i(x) | i \in I_\delta(x)\} \\ c_2(x) &= \max_i\{\|f_i'(x)\| | i \in I_\delta(x)\}. \end{aligned}$$

Thus, if β and p satisfy the constraints of problem (3.8), then

$$\beta + 1/2(p, Ap) \geq c_1(x) - \|p\| c_2(x) + m\|p\|^2.$$

This bound shows that if $\beta \to \infty$ and $p \to \infty$, then the goal function of problem (3.8) tends to $+\infty$. Thus, it is easy to deduce that the minimum of problem (3.8) is attained. Moreover, if we use the fact that for fixed p the minimum of (3.8) with respect to β is attained for

$$\beta = \max_i \{(f_i'(x), p) + f_i(x) | i \in I_\delta(x)\}$$

then problem (3.8) may be rewritten in the form

$$\min_p \{1/2(p, Ap) + \max_i \{(f_i'(x), p) + f_i(x) | i \in I_\delta(x)\}\}$$

which is equivalent to the original. By virtue of (3.7), this problem comprises the minimization of a convex function [23] and thus its minimum is attained at a unique point.

All the above enables us to formulate the following theorem.

Theorem 3.1 *If the condition (3.7) is satisfied, then problem (3.8) has a unique minimum $p_A(x)$, $\beta_A(x)$ and*

$$\beta_A(x) = \max_i \{(f_i'(x), p_A(x)) + f_i(x) | i \in I_\delta(x)\}.$$

In order to derive necessary conditions for an extremum and to formulate the dual problem, we construct the Lagrange function for problem (3.8). According to Section 1.2.3

$$\begin{aligned} L(p, \beta, u) &= \beta + 1/2(p, Ap) + \sum_{i \in I_\delta(x)} u^i[(f_i'(x), p) + f_i(x) - \beta] \\ &= [1 - \sum_{i \in I_\delta(x)} u^i]\beta + 1/2(p, Ap) \\ &\quad + (\sum_{i \in I_\delta(x)} f_i'(x)u^i, p) + \sum_{i \in I_\delta(x)} u^i f_i(x). \end{aligned} \qquad (3.9)$$

According to the above and Theorems 2.9 and 2.6, for the minimum $p = p_A(x)$, $\beta = \beta_A(x)$, there exist multipliers $u_A'(x)$ such that

$$u_A^i(x) \geq 0, \quad u_A^i(x)[(f_i'(x), p_A(x)) + f_i(x) - \beta_A(x)] = 0 \qquad (3.10)$$

and, substituting $u_A^i(x)$ in (3.9), $L(p, \beta, u_A(x))$ attains its minimum over $p \in \mathbb{R}^n$, $\beta \in \mathbb{R}^1$ for $p = p_A(x)$, $\beta = \beta_A(x)$. Differentiating (3.9) with respect to β and p and equating the derivatives to zero, we obtain

$$\sum_{i \in I_\delta(x)} u_A^i(x) = 1, \quad Ap_A(x) + \sum_{i \in I_\delta(x)} u_A^i(x)(f_i'(x))^* = 0. \qquad (3.11)$$

Thus, equations (3.10) and (3.11) are necessary and sufficient conditions which link the minimum and the Lagrange multipliers of problem (3.8).

To construct the dual problem corresponding to (3.8), according to Section 1.2.4, we have to compute the minimum in (3.9) with respect to p and β for fixed u^i, $i \in I_\delta(x)$, $u^i \geq 0$. But

$$\begin{aligned}\varphi(u) &= \min_{p,\beta} L(p,\beta,u) \\ &= \begin{cases} -\infty & \sum_{i\in I_\delta(x)} u^i \neq 1 \\ -\frac{1}{2}(\sum_{i\in I_\delta(x)} u^i f_i'(x), \sum_{i\in I_\delta(x)} u^i A^{-1}(f_i'(x))^*) & \\ \quad + \sum_{i\in I_\delta(x)} u^i f_i(x) & \sum_{i\in I_\delta(x)} u^i = 1 \end{cases}\end{aligned}$$

Thus, according to Section 1.2.4, the dual problem corresponding to (3.8) has the form

$$\max_{u\geq 0}\{-\frac{1}{2}(\sum_{i\in I_\delta(x)} u^i f_i'(x), \sum_{i\in I_\delta(x)} u^i A^{-1}(f_i'(x))^*) + \sum_{i\in I_\delta(x)} u^i f_i(x) | \sum_{i\in I_\delta(x)} u^i = 1\}. \tag{3.12}$$

Here, the $u_A^i(x)$ are the solution of problem (3.12). The uniqueness arguments for $p_A(x)$ and $\beta_A(x)$ used in the proof of Theorem 1.17 show that $p_A(x)$ may be expressed in terms of $u_A(x)$ using formula (3.11), i.e.

$$p_A(x) = -\sum_{i\in I_\delta(x)} u_A^i(x) A^{-1}(f_i'(x))^*. \tag{3.13}$$

3.1.2 Some Bounds

The subsequent construction of the algorithm is based on a number of bounds on the variation of the functions $f_i(x)$ for displacements in directions determined by the solution of problem (3.8). By scalar multiplication of the second equation of (3.11) by $p_A(x)$, taking into account the first equation and (3.10). we obtain

$$(p_A(x), Ap_A(x)) - \sum_{i\in I_\delta(x)} u^i f_i(x) + \beta_A(x) = 0.$$

It follows from this equation, together with (3.11), that

$$(p_A(x), Ap_A(x)) \leq F(x) - \beta_A(x) \tag{3.14}$$

since $f_i(x) \leq F(x)$.

Suppose now that x_0 is fixed, that $\alpha = F(x_0)$ and that in the region C_α

$$K = \max_{x\in C_\alpha, i\in I} \|f_i'(x)\|.$$

We also suppose that L is the Lipschitz constant for the gradients in the region C_α, i.e.

$$\|f_i'(x_1) - f_i'(x_2)\| \leq L\|x_1 - x_2\|, \quad x_1, x_2 \in C_\alpha.$$

Using the well-known mean value formula from mathematical analysis, we obtain that for some θ, $0 \leq \theta \leq 1$

$$\begin{aligned} f_i(x+\alpha p) &= f_i(x)+\alpha(f_i'(x+\theta\alpha p),p) \\ &= f_i(x)+\alpha(f_i'(x),p)+\alpha(f_i'(x+\theta\alpha p)-f_i'(x),p) \\ &\le f_i(x)+\alpha(f_i'(x),p)+\alpha^2L\|p\|^2. \end{aligned} \tag{3.15}$$

Suppose that $i \in I_\delta(x)$ and

$$p = p_A(x). \tag{3.16}$$

Then, taking into account the fact that p satisfies the constraints of problem (3.8) together with the inequality (3.14), we obtain that for $0 \le \alpha \le 1$

$$\begin{aligned} f_i(x+\alpha p) &\le f_i(x)+\alpha(\beta_A(x)-f_i(x))+\alpha^2L\|p\|^2 \\ &= (1-\alpha)f_i(x)+\alpha\beta_A(x)+\alpha^2L\|p\|^2 \\ &\le (1-\alpha)F(x)+\alpha\beta_A(x)+\alpha^2L\|p\|^2 \\ &= F(x)-\alpha(F(x)-\beta_A(x))+\alpha^2L\|p\|^2 \\ &\le F(x)-\alpha(p,Ap)+\alpha^2L\|p\|^2. \end{aligned}$$

Thus, for $i \in I_\delta(x)$ and $p = p_A(x)$ the following inequality holds:

$$f_i(x+\alpha p) \le F(x)-\alpha(p,Ap)+\alpha^2L\|p\|^2, \quad 0 \le \alpha \le 1. \tag{3.17}$$

Suppose now that $i \notin I_\delta(x)$. Then

$$f_i(x+\alpha p) = f_i(x)+\alpha(f_i'(x+\theta\alpha p),p) \le F(x)-\delta+\alpha K\|p\|. \tag{3.18}$$

But the inequality

$$F(x)-\alpha(p,Ap) \ge F(x)-\delta+\alpha K\|p\| \tag{3.19}$$

holds if

$$\alpha \le \frac{\delta}{(p,Ap)+K\|p\|}.$$

Thus, comparing (3.17)–(3.19), we obtain

$$F(x+\alpha p) \le F(x)-\alpha(p,Ap)+\alpha^2L\|p\|^2 \tag{3.20}$$

for

$$0 \le \alpha \le \min\left[1, \frac{\delta}{(p,Ap)+K\|p\|}\right]. \tag{3.21}$$

Suppose that $0 < \epsilon < 1$. Then

$$(p,Ap)-\alpha L\|p\|^2 \ge \epsilon(p,Ap) \tag{3.22}$$

for

$$0 \le \alpha \le \frac{1-\epsilon}{L}\frac{(p,Ap)}{\|p\|^2}. \tag{3.23}$$

Comparing (3.20)–(3.23), we see that the following result holds.

Lemma 3.1 *The bound*

$$F(x + \alpha p_A(x)) \leq F(x) - \alpha\epsilon(p_A(x)), Ap_A(x)) \tag{3.24}$$

holds for the parameter α *satisfying the inequality*

$$0 \leq \alpha \leq \min\left[1, \frac{\delta}{(p, Ap) + K\|p\|}, \frac{1-\epsilon}{L}\frac{(p, Ap)}{\|p\|^2}\right].$$

We also note the following:

Lemma 3.2 *The function* $p_A(x)$ *is equal to zero if and only if the necessary conditions for a minimum of the function* $F(x)$ *are satisfied at the point* x.

Proof. Clearly, if $p_A(x) = 0$, then $\beta_A(x) = F(x)$ and equations (3.10) and (3.11) lead to the equations of (3.6). Conversely, if the equations of (3.6) hold, then for $p_A(x) = 0$ and $\beta_A(x) = F(x)$, they become the equations (3.10) and (3.11) which are sufficient conditions for $p_A(x)$ and $\beta_A(x)$ to be a solution of problem (3.8). □

3.1.3 Algorithms

We now formulate the linearization algorithm for the minimax problem.

Suppose that the numbers $\delta > 0$ and $0 < \epsilon < 1$ and an initial point x_0 have been chosen.

General step: if the approximation x_k has already been constructed then we choose a symmetric, strictly positive-definite matrix A_k.

1. We solve problem (3.8) for $x = x_k$ and $A = A_k$ and set

$$p_k = p_{A_k}(x_k), \quad u_k = u_{A_k}(x_k).$$

2. We choose the step α_k by dividing unity by two until the following inequality is satisfied.

$$F(x_k + \alpha_k p_k) \leq F(x_k) - \alpha_k\epsilon(p_k, A_k p_k). \tag{3.25}$$

3. We set $x_{k+1} = x_k + \alpha_k p_k$ and return to stage 1.

   Stopping condition: $p_k = 0$ or $\|p_k\| \leq \epsilon_1$, where ϵ_1 is the accuracy, which may be assigned.

The following theorem gives conditions for convergence of the algorithm.

Theorem 3.2 *If*

$$\|A_k\| \leq M, \quad (p, A_k p) \geq m\|p\|^2$$

for all k, $p \in \mathbb{R}^n$, then any limit point x_* of the sequence $\{x_k\}$ satisfies the necessary conditions for an extremum (3.6) and $\|p_k\| \to 0$.

Proof. The algorithm generates the sequence of points x_k through which the function $F(x)$ decreases. Thus, all the points x_k belong to the compact set

$$C_\alpha = \{x | F(x) \leq \alpha\}, \quad \alpha = F(x_0).$$

It follows from (3.11) that

$$\begin{aligned} \|p_k\| &= \| \sum_{i \in I_\delta(x_k)} u_k^i A_k^{-1} (f_i'(x_k))^* \| \\ &\leq \max_{i \in I} \|A_k^{-1}\| \|f_i'(x_k)\| \leq K/m. \end{aligned}$$

Thus, the vectors p_k have bounded norms. From Lemma 3.1, if we coarsen the bound on the value of the step α, we obtain

$$F(x_k + \alpha p_k) \leq F(x_k) - \alpha\epsilon(p_k, A_k p_k)$$

if $0 \leq \alpha \leq \alpha_0$, where

$$\alpha_0 = \min\left[1, \frac{\delta}{M(K/m)^2 + K^2/m}, \frac{1-\epsilon}{L} m\right] > 0.$$

Whence it follows that the step α_k obtained by dividing unity by two until the inequality (3.25) is satisfied cannot be less than half of α_0, i.e. $\alpha_k \geq \alpha_0/2$. Since C_α is compact, we have bounded the function $F(x)$ below. Thus, from (3.25) it follows that

$$\alpha_k(p_k, A_k p_k) \to 0$$

or, since $\alpha_k \geq 0.5\alpha_0$,

$$(p_k, A_k p_k) \geq m\|p_k\|^2 \to 0.$$

Thus, $p_k \to 0$.

Suppose now that $J \subseteq \{0, 1, \ldots\}$ is any sequence of indices k, through which $x_k \to x_*$. Because the set C_α is compact, such a point x_* and sequence J always exist. Since the coefficients u_k^i are nonnegative and their sum is equal to 1, then we may assume, without loss of generality, that $u_k^i \to u_*^i$, $k \in J$ and

$$u_*^i \geq 0, \quad \sum_{i=1}^m u_*^i = 1.$$

Here, we have assumed that $u_k^i = 0$, $i \notin I_\delta(x_k)$; thus, the coefficients u_k^i are defined for all $i \in I$.

Moreover, $\beta_k = \beta_{A_k}(x_k) \leq F(x_k)$, by virtue of (3.14). On the other hand,

$$\beta_k \geq f_i(x_k) + (f_i'(x_k), p_k) \geq f_i(x_k) - K\|p_k\|, \quad i \in I_\delta(x_k).$$

Thus,

$$F(x_k) \geq \beta_k \geq F(x_k) - K\|p_k\|$$

and since $\|p_k\| \to 0$ we have

$$\beta_k \to F(x_*).$$

By virtue of (3.10), (3.11) and the fact that, by convention, $u_k^i = 0$, $i \notin I_\delta(x_k)$,

$$\begin{aligned} u_k^i \geq 0,\ \ u_k^i[(f_i'(x_k), p_k) + f_i(x_k) - \beta_k] &= 0,\ \ i \in I \\ A_k p_k + \sum_{i=1}^m u_k^i (f_i'(x_k))^* &= 0 \\ \sum_{i=1}^m u_k^i &= 1. \end{aligned}$$

Then, letting $k \in J$ tend to infinity, the equations of (3.6) follow immediately from the above equations, since $p_k \to 0$, $u_k \to u_*$ and $\beta_k \to F(x_*)$. The proof of the theorem is complete. □

3.1.4 Algorithm for $A_k = I_n$

The simplest algorithm to implement is one in which all the matrices A_k are equal to the unit matrix. In fact, in this case, the dual problem of the auxiliary problem (3.8) has the simple form

$$\begin{aligned} \max_{u \geq 0} \{ -\frac{1}{2} \| \sum_{i \in I_\delta(x)} u^i (f_i'(x))^* \|^2 + \sum_{i \in I_\delta(x)} u^i f_i(x) : \\ \sum_{i \in I_\delta(x)} u^i = 1,\ u^i \geq 0 \} \end{aligned} \tag{3.26}$$

and the solution of the original problem is given in terms of the solution of the dual problem by the simple formula

$$p(x) = - \sum_{i \in I_\delta(x)} u^i (f_i'(x))^*.$$

In what follows, for $A = I_n$, we shall simply denote $p_A(x)$ and $u_A(x)$ by $p(x)$ and $u(x)$.

Let us consider the rate of convergence of the algorithm for $A_k = I_n$ in more detail. For this, we require a number of additional assumptions.

1. Suppose that x_* is the unique minimum of $F(x)$ and is the only point at which the necessary conditions for a minimum (3.6) are satisfied

2. The functions $f_i(x)$, $i \in I$ are twice continuously differentiable.

3. If

$$I_* = I_0(x_*) = \{i \in I | f_i(x_*) = F(x_*)\}$$

then $u_*^i > 0$, $i \in I_*$ and the vectors $f_i'(x_*) - f_j'(x_*)$, $i \in I_*$, $i \neq j$, where j denotes any index in I_*, are linearly independent.

4. $(p, L''_{xx}(x_*, u_*)p) > 0$ for all $p \neq 0$, where

$$L(x, u) = \sum_{i=1}^{m} u^i f_i(x)$$

and L''_{xx} is the matrix of second derivatives of L with respect to x.

Everything that follows in this section is now analogous to Section 2.3 and thus we may shorten the proofs somewhat.

Lemma 3.3 *There exists a neighbourhood of the point x_* in which $I_* \subseteq I_\delta(x)$.*

Proof. The proof is analogous to that of Lemma 2.2. □

Lemma 3.4 *In some neighbourhood of the point x_*, the functions $p(x)$ and $u(x)$ are uniquely defined by the system of equations*

$$p(x) + \sum_{i \in I_*} u^i(x)(f_i'(x))^* = 0 \tag{3.27}$$

$$(f_i'(x), p(x)) + f_i(x) = \beta(x), \quad i \in I_* \tag{3.28}$$

$$\sum_{i \in I_*} u^i(x) = 1 \tag{3.29}$$

and are differentiable functions of x.

Proof. Clearly, for $x = x_*$, by virtue of (3.6), one solution of the system (3.27) is $p(x_*) = 0$, $u^i(x_*) = u_*^i$, $\beta(x_*) = F(x_*)$.

Suppose now that j is an arbitrary index in I_* and

$$\begin{aligned} \tilde{I}_* &= I_* \backslash \{j\} \\ \tilde{f}_i(x) &= f_i(x) - f_j(x), \quad i \in I_*, \quad i \neq j, \quad \tilde{f}_0(x) = f_j(x). \end{aligned}$$

Then, subtracting equation (3.28) for the index j from the other equations of (3.28) and carrying out obvious transformations of (3.27) and (3.29), we see that the system (3.27)–(3.29) is equivalent to the system

$$\begin{aligned} p(x) + (\tilde{f}_0'(x))^* + \sum_{i \in \tilde{I}_*} u^i(x)(\tilde{f}_i'(x))^* &= 0 \\ (\tilde{f}_i'(x), p(x)) + \tilde{f}_i(x) &= 0, \quad i \in \tilde{I}_* \\ u_j(x) &= 1 - \sum_{i \in \tilde{I}_*} u^i(x). \end{aligned} \tag{3.30}$$

But the system (3.30) is completely analogous to the system (2.95), thus, arguing as in the proof of Lemma 2.3, we conclude that the system (3.30) (whence also (3.27)–(3.29)) has a unique solution in a neighbourhood of x_*, which is continuously differentiable and that $u^i(x) \geq 0$ for $i \in I_*$. Equating (3.27)–(3.29) with the necessary and sufficient conditions for a minimum (3.10) and (3.11) and setting $u^i(x) = 0$, $i \notin I_*$, we deduce that in the chosen neighbourhood the solution of the system (3.27)–(3.29) is also a solution of the auxiliary problem (3.8). □

Since the systems (3.30) and (2.95) are analogous, the properties of the eigenvalues of the matrix $p'(x_*)$ may be established and thus we can formulate a theorem and corollary analogous to Theorem 2.14 and its Corollary.

Theorem 3.3 *If assumptions 1–4 of this section are satisfied then there exist a neighbourhood of the point x_* and a number $\alpha > 0$ such that the process*

$$x_{k+1} = x_k + \alpha p(x_k)$$

converges to x_* with the rate of a geometric progression with denominator less than one.

Corollary *If assumptions 1–3 are satisfied and the set I_* contains $n+1$ indices, then the process*

$$x_{k+1} = x_k + p(x_k)$$

converges superlinearly in some neighbourhood of the point x_*.

The Corollary requires some explanation. The set $\widetilde{I}_*$ contains one index less than the set I_*. In the Corollary to Theorem 2.14 there is a set which corresponds to the set $\widetilde{I}_*$ here. Thus, for the set $\widetilde{I}_*$ to contain n indices, the set I_* must have order $n+1$.

We now note that the situation described in the Corollary to Theorem 2.14 is quite typical in minimax problems, in particular in problems of approximation theory (more details of this are given in [6,7]). Thus, it is desirable to show that, in this situation, the linearization algorithm with $A_k = I_n$ guarantees a unit step in a neighbourhood of the point x_* and thus also superlinear convergence. The fact that this is actually the case was proved in [4] and we shall follow this.

Lemma 3.5 *Suppose that assumptions 1–3 of this section are satisfied and that $|I_*| = n+1$. Then there exist numbers $r > 0$ and $\eta > 0$ such that*

$$\min_{\|p\|=1} \max_{i \in I_*} (f_i'(x), p) \geq \eta$$

for all x such that $\|x - x_*\| \leq r$.

Proof. We shall show that

$$\max_{i \in I_*} (f_i'(x_*), p) > 0$$

for any p with $\|p\| = 1$. Clearly, if this is not the case, then there exists a vector p such that

$$(f_i'(x_*), p) \leq 0, \quad i \in I_*. \tag{3.31}$$

But then, from (3.6) and the assumptions 1–3, it follows that

$$\sum_{i \in I_*} u_*^i f_i'(x_*) = 0, \quad u_*^i > 0, \quad i \in I_*.$$

Scalar multiplication of the first equation by p gives

$$\sum_{i \in I_*} u_*^i (f_i'(x_*), p) = 0. \tag{3.32}$$

Since, by virtue of (3.31), all the terms of this are non-positive, (3.32) is only possible if

$$(f_i'(x_*), p) = 0, \quad i \in I_*$$

or

$$(f_i'(x_*) - f_j'(x_*), p) = 0, \quad i \in \tilde{I}_*.$$

Whence, we deduce that the vector p is orthogonal to n linearly independent vectors and so $p = 0$, in contradiction to the fact that $\|p\| = 1$.

We now note that the functions

$$\begin{aligned} \varphi_0(x, p) &= \max_{i \in I_*}(f_i'(x), p) \\ \varphi_1(x) &= \min_{\|p\|=1} \varphi_0(x, p) \end{aligned}$$

are continuous in x and p. This was proved, for example, in [6,23]. Thus, from the above, it follows that

$$\varphi_0(x_*, p) > 0, \quad \|p\| = 1, \quad \varphi_1(x_*) > 0$$

and there exists a neighbourhood of the point x_* such that

$$\varphi_1(x) \geq 1/2\varphi_1(x_*).$$

Setting r equal to the radius of this neighbourhood and $\eta = 1/2\varphi_1(x_*)$, we obtain the required result. □

We now note that if the point x is sufficiently close to x_* then $p(x)$ is close to $p(x_*) = 0$. Thus, the point $x + p(x)$ is also close to x_*. Whence, it is easy to show that in some neighbourhood of the point x_* there always exists an index $i \in I_*$ (it may depend on x) such that

$$F(x + p(x)) = f_i(x + p(x)).$$

We choose this neighbourhood sufficiently small that the inequality

$$\max_{i \in I_*} \|f_i'(x + \theta_i p(x)) - f_i'(x)\| \leq (1 - \epsilon)\eta$$

is satisfied in it for any θ_i, $0 \leq \theta_i \leq 1$, where $0 < \epsilon < 1$. Then, using (3.28), we obtain

$$\begin{aligned}
F(x) - \beta(x) &\geq \max_{i \in I_*}(f_i(x) - \beta(x)) \\
&= \max_{i \in I_*}(f_i'(x), -p(x)) \geq \eta \|p(x)\| \\
F(x + p(x)) - \beta(x) &= \max_{i \in I_*}[f_i(x + p(x)) - f_i(x) - (f_i'(x), p(x))] \\
&= \max_{i \in I_*}(f_i'(x + \theta_i p(x)) - f_i'(x), p(x)) \\
&\leq (1 - \epsilon)\eta \|p(x)\| \leq (1 - \epsilon)[F(x) - \beta(x)] \qquad (3.33)
\end{aligned}$$

Since from (3.14) it follows that

$$\|p(x)\|^2 \leq F(x) - \beta(x)$$

because $A = I_n$, then

$$\begin{aligned}
F(x) - F(x + p(x)) &= F(x) - \beta(x) - [F(x + p(x)) - \beta(x)] \\
&\geq \epsilon[F(x) - \beta(x)] \geq \epsilon \|p(x)\|^2.
\end{aligned}$$

Thus, we have proved the following lemma.

Lemma 3.6 *Under the assumptions of Lemma 3.5, there exists a neighbourhood of the point x_* in which the inequality*

$$F(x + p(x)) \leq F(x) - \epsilon \|p(x)\|^2$$

is satisfied.

Now we may establish the main result.

Theorem 3.4 *Suppose that assumptions 1–3 are satisfied and that $|I_*| = n + 1$. Then the linearization algorithm with $A_k = I_n$ converges superlinearly (i.e. faster than any geometric progression) to the minimum x_* of the function $F(x)$.*

Proof. Since the point x_* is the only point at which the necessary conditions for a minimum of the function $F(x)$ are satisfied, then, according to Theorem 3.2, the sequence generated by the algorithm has a unique limit point x_* and thus $x_k \to x_*$. According to the algorithm, the step α_k is chosen by dividing one by two and using inequality (3.25), which now takes the form

$$\begin{aligned}
F(x_k + \alpha_k p_k) &\leq F(x_k) - \alpha_k \epsilon \|p_k\|^2 \\
p_k &= p(x_k).
\end{aligned}$$

But Lemma 3.6 shows that, for k sufficiently large, when x_k is close to x_*, this inequality will be satisfied for $\alpha_k = 1$, since for large k, the formula

$$x_{k+1} = x_k + p(x_k)$$

holds. The proof may now be completed by applying Theorem 3.3. □

3.1.5 Acceleration of Convergence in the Convex Case

In our study of the rate of convergence, Lemma 3.4, which assigns the system of equations which $p(x)$ and $u(x)$ satisfy in a neighbourhood of the minimum of $F(x)$ played a crucial role. Here, the vector $p(x)$ corresponded to the matrix $A = I_n$. However, it is easy to see that if we take an arbitrary positive definite matrix A then $p_A(x)$ will satisfy a system analogous to (3.27)–(3.29) in a neighbourhood of x_*:

$$\begin{aligned} Ap_A(x) + \sum_{i \in I_*} u_A^i(x)(f_i'(x))^* &= 0 \\ (f_i'(x), p_A(x)) + f_i(x) &= \beta_A(x), \quad i \in I_* \\ \sum_{i \in I_*} u_A^i(x) &= 1 \end{aligned} \tag{3.34}$$

This system may be obtained directly from (3.10) and (3.11) provided assumptions 3 and 4 are satisfied. Here, as in Lemma 3.4, $p_A(x)$ and $u_A^i(x)$ are continuously differentiable in a neighbourhood of the point x_*. In fact, the system (3.34) is linear in p_A, u_A and β_A, where the matrix of this system at the point $x = x_*$ is nonsingular (we leave this to the reader to verify, using assumption 3). Thus, $p_A(x)$, $u_A(x)$ and $\beta_A(x)$ are uniquely expressed in terms of the parameters of problem (3.34) and if the $f_i(x)$ are twice continuously differentiable, then p_A, u_A and β_A are continuously differentiable.

Suppose that $u_A^i(x) = 0$, $i \notin I_*$ and that $f'(x)$ denotes the matrix with rows $f_i'(x)$, $i \in I_*$. Differentiating (3.34) with respect to x for $x = x_*$ and taking into account the fact that $p_A(x_*) = 0$, $u_A(x_*) = u_*$ and $\beta_A(x_*) = F(x_*)$, we obtain

$$\begin{aligned} Ap_A'(x_*) + L_{xx}''(x_*, u_*) + \sum_{i \in I_*} (f_i'(x_*))^*(u_A^i(x_*))' &= 0 \\ f'(x_*)p_A'(x_*) + f'(x_*) &= \mathbf{1}\beta_A'(x_*) \\ \sum_{i \in I_*} (u_A^i(x_*))' &= 0 \end{aligned} \tag{3.35}$$

where $\mathbf{1}$ is the vector of length $|I_*|$ all the components of which are equal to one.

Since the matrix of the system (3.34) was nonsingular, then according to the implicit function theorem [23], the system (3.35) for the derivatives is also solvable. Suppose that $A = L_{xx}''(x_*, u_*)$ and that assumption 4 of Section 3.1.4 is satisfied. For this A.

$$p_A'(x_*) = -I_n, \quad (u_A^i(x_*))' = 0, \quad \beta_A'(x_*) = 0 \tag{3.36}$$

is a solution of the system (3.35).

Thus, we have shown that the following result holds.

Theorem 3.5 *Suppose that assumptions 1–4 of Section 3.1.4 are satisfied. Then for $A = L_{xx}''(x_*, u_*)$ the solution of the dual problem (3.8) is differentiable in some neighbourhood of the point x_* and formula (3.36) holds.*

The validity of formula (3.36) enables us to formulate the following algorithm for which we may expect a higher rate of convergence.

Algorithm Suppose that x_0, $\epsilon \in (0,1)$, $\gamma \in (0,1)$ and $\delta > 0$ have been chosen. We set $C_0 = +\infty$.

General step: if the points x_k and u_{k-1} and the number C_k have already been constructed then we proceed as follows

1. We set $A_k = L''_{xx}(x_k, u_{k-1})$, $k \geq 1$, $A_0 = I_n$.

2. We solve problem (3.8) with $x = x_k$ and $A = A_k$ and calculate $p_k = p_{A_k}(x_k)$ and $u_k = u_{A_k}(x_k)$.

3. If $\|p_k\| \leq C_k$ then we set

$$\overline{x} = x_k + p_k, \quad A = L''_{xx}(\overline{x}, u_k)$$

and calculate the vector $p_A(\overline{x})$.

If $\|p_A(\overline{x})\| \leq \gamma\|p_k\|$, then $x_{k+1} = \overline{x}$ and $C_{k+1} = \gamma\|p_k\|$.

If $\|p_A(\overline{x})\| > \gamma\|p_k\|$ then we set $C_{k+1} = \gamma\|p_k\|$ and move to stage 5.

4. If $\|p_k\| > C_k$, then $C_{k+1} = C_k$ and we move to stage 5.

5. Beginning with $\alpha = 1$, we divide this value by two until the inequality

$$F(x_k + \alpha_k p_k) \leq F(x_k) - \epsilon\alpha_k(p_k, A_k p_k)$$

is satisfied.

We set

$$x_{k+1} = x_k + \alpha_k p_k$$

and return to stage 1.

Theorem 3.6 *Suppose that $f_i(x)$, $i \in I$ are twice continuously differentiable convex functions and there exist constants $M \geq m > 0$ such that*

$$m\|p\|^2 \leq (f_i''(x)p, p) \leq M\|p\|^2, \quad i \in I \tag{3.37}$$

Suppose that x_* is the unique minimum of the function $F(x)$, that the vectors $f_i'(x_*) - f_j'(x_*)$, $i \in I_* - \{j\}$, where j is any index in I_*, are linearly independent and that $u_*^i > 0$ for $i \in I_*$, Then the given algorithm generates a sequence x_k which converges to x_* and

$$\|x_{k+1} - x_*\| \leq q_k\|x_k - x_*\|, \quad q_k \to 0.$$

Proof. The proof of this theorem is largely a restatement of the proof of Theorems 2.15 and 2.17.

However, a number of comments are in order at this point. By virtue of condition (3.37), the function $F(x)$ is strictly convex and its minimum is attained at a unique point x_*. If $u^i \geq 0$ and the sum of the u^i is equal to one, then it follows from (3.37) that

$$m\|p\|^2 \leq (L''_{xx}(x,u)p,p) \leq M\|p\|^2.$$

Thus, all the matrices $A_k = L''_{xx}(x_k, u_{k-1})$ satisfy the conditions of Theorem 3.2. If this algorithm did not include stages 3 and 4, then it would simply be the same as the linearization method and, according to Theorem 3.2, would generate a sequence converging to x_*.

Proceeding now exactly as in the proof of Theorem 2.17, we deduce that there exists an infinite sequence of indices J^0, such that

$$x_k \to x_*, \quad u_{k-1} \to u_*, \quad A_k = L''_{xx}(x_k, u_{k-1}) \to L''_{xx}(x_*, u_*) = A.$$

Suppose that $p_0(x)$ and $u_0^i(x)$ represent the solution of (3.8) for $A = L''_{xx}(x_*, u_*)$, which, as previously shown, satisfies the system (3.34) with corresponding matrix A. On the other hand, p_k and u_k satisfy the system (3.34) for $A = A_k$ and $x = x_k$, for k sufficiently large, $k \in J^0$. Subtracting these systems from one another, we obtain

$$\begin{aligned} A_k(p_k - p_0(x_k)) + \sum_{i\in I_*}(u_k^i - u_0^i(x_k))(f_i'(x_*))^* &= -(A_k - A)p_0(x_k) \\ f'(x_k)(p_k - p_0(x_k)) - (\beta_k - \beta_0(x_k)) &= 0 \\ \sum_{i\in I_*}(u_k^i - u_0^i(x_k)) &= 0. \end{aligned}$$

The matrix of this system of equations in the differences is the same as that for the system (3.34), i.e. it is nonsingular and the right-hand side is equal to $(A_k - A)p_0(x_k)$. Thus, solving this system in the differences $p_k - p_0(x_k)$, $u_k - u_0(x_k)$, $\beta_k - \beta_0(x_k)$, we obtain

$$\begin{aligned} \|p_k - p_0(x_k)\| &\leq C\|A_k - A\|\|p_0(x_k)\| \\ \|u_k - u_0(x_k)\| &\leq C\|A_k - A\|\|p_0(x_k)\| \\ |\beta_k - \beta_0(x_k)| &\leq C\|A_k - A\|\|p_0(x_k)\|. \end{aligned}$$

We now recall that, according to the equations of (3.36),

$$\begin{aligned} p_0(x) &= -(x - x_*) + \|x - x_*\|\omega_0(x, x_*) \\ u_0(x) &= u_* + \|x - x_*\|\omega_1(x, x_*) \\ \beta_0(x) &= F(x_*) + \|x - x_*\|\omega_2(x, x_*) \\ \lim_{x\to x_*} \omega_j(x, x_*) &= 0, \quad j = 0, 1, 2. \end{aligned} \tag{3.38}$$

Therefore, if $k \in J^0$, then

$$\begin{aligned} \overline{x} &= x_k + p_k = x_k + p_0(x_k) + (p_k - p_0(x_k)) \\ &= x_* + \|x_k - x_*\|\omega_0(x_k, x_*) + (p_k - p_0(x_k)) \\ \|\overline{x} - x_*\| &\leq \|x_k - x_*\|\omega_0(x_k, x_*) + C\|A_k - A\|\|p_0(x_k)\| \\ &\leq [\omega_0(x_k, x_*) + C\|A_k - A\|]C_1\|x_k - x_*\| \end{aligned} \tag{3.39}$$

where we have used the fact that $p_0(x)$ and $\|x - x_*\|$ are values of the same order of magnitude. In the same way, it follows from (3.38) and (3.39) that

$$\|p_0(\overline{x})\| \leq [\omega_0(x_k, x_*) + C\|A_k - A\|]C_2\|p_0(x_k)\|.$$

Then, taking into account the bound on the distance between p_k and $p_0(x_k)$ and, by virtue of the bounds

$$\begin{aligned} \|u_k - u_*\| &\leq \|x_k - x_*\|\omega_1(x_k, x_*) + C\|A_k - A\|\|p_0(x_k)\| \\ &\leq [\omega_1(x_k, x_*) + C_3\|A_k - A\|]C_4\|x_k - x_*\| \end{aligned}$$

we deduce that

$$\|p_{L''_{xx}(\overline{x}, u_k)}(\overline{x})\| \leq g_k\|p_k\|$$

where $g_k \to 0$, $k \in J^0$.

Thus $g_k < \gamma$ for large k and stage 3 of the algorithm is successful. Therefore, for large $k \in J^0$, the algorithm step will be successful and $x_{k+1} = \overline{x}$ and u_k will be much closer to x_* and u_*, respectively. Repeating the above reasoning, we deduce that at the point x_{k+1}, stage 3 of the algorithm will be successful. Thus, it follows that, from some instant onwards, the algorithm will operate according to the formula $x_{k+1} = x_k + p_k$ and the bound (3.39) will hold:

$$\begin{aligned} \|x_{k+1} - x_*\| &\leq q_k\|x_k - x_*\| \\ q_k &= C_1[\omega_0(x_k, x_*) + C\|A_k - A\|]. \end{aligned}$$

But $\omega_0(x_k, x_*) \to 0$ and $A_k \to A$, thus $q_k \to 0$ and the proof of the theorem is complete. □

3.2 The Dual Algorithm for Convex Programming Problems

In recent years, the method of modified Lagrange multipliers has achieved a broad popularity. There is a considerable literature on this topic [21,34]. In this section, we shall describe and use a method for solving convex programming problems which is very similar to the method of modified Lagrange functions. However, the properties of this method are studied by techniques different from those used in the literature referred to above. Thus, this method has all the inherent advantages of the method of modified Lagrange functions while the rules for good convergence of the method and techniques for accelerating its convergence are absolutely explicit.

Suppose that $f_i(x)$, $i = 0, 1, \ldots, m$ are convex functions on $\mathbb{R}^n$, defined on a domain containing the convex set $X \subseteq \mathbb{R}^n$. The problem

$$\min_x\{f_0(x) | f_i(x) \leq 0,\ i = 1, \ldots, m,\ x \in X\} \tag{3.40}$$

is a convex programming problem. Let us consider the subset M of $\mathbb{R}^{m+1}$ given by

$$M = \{z \in \mathbb{R}^{m+1} | z = f(x) + v,\ x \in X,\ v \geq 0\} = f(X) + \mathbb{R}_+^{m+1} \tag{3.41}$$

where the vector z has components $z^0, z^1, \ldots, z^m$, $f(x)$ is the vector with components $f_i(x)$, $i = 0, 1, \ldots, m$ and $\mathbb{R}_+^{m+1}$, as usual, denotes the positive orthant in $\mathbb{R}^{m+1}$.

It is easy to see that M is a convex set and that the convex programming problem (3.40) is equivalent to the problem

$$\min\{\lambda | \lambda e \in M\} \tag{3.42}$$

where e is the vector with components $e^0 = 1$, $e^j = 0$, $i = 1, \ldots, m$. Thus, in the next paragraphs we shall concentrate on solving problem (3.42). We denote its solution by λ_*, so that $\lambda e \notin M$, $\lambda < \lambda_*$, $\lambda_* e \in M$.

Let us now move away from the specific form (3.41) of the set M. We shall assume that M is a closed convex set and that the vector e is an arbitrary unit vector.

3.2.1 The Dual Algorithm

Let us formulate the algorithm for solving problem (3.42). Suppose that

$$\lambda_0 < \lambda_*,\quad y_0 = \lambda_0 e,\quad z_0 = \text{argmin}\,\{\|z - y_0\| | z \in M\}.$$

We also set

$$\begin{aligned} r_0 &= \|z_0 - y_0\| \\ \eta_0 &= (z_0 - y_0)/r_0,\quad \|\eta_0\| = 1 \\ \lambda_1 &= (\eta_0, z_0)/(\eta_0, e). \end{aligned}$$

It is known (see Section 1.2.1) that the inequality

$$(z, \eta_0) \geq (z_0, \eta_0),\quad z \in M$$

is satisfied.

Let us now define a general step of the algorithm. Suppose that $\lambda_k \leq \lambda_*$, η_{k-1} and z_{k-1} have already been constructed and satisfy the equations

$$\lambda_k(\eta_{k-1}, e) = (\eta_{k-1}, z_{k-1}) \leq (\eta_{k-1}, z),\quad z \in M,\ z_{k-1} \in M,\ \|\eta_{k-1}\| = 1. \tag{3.43}$$

We set

$$\begin{aligned} y_k &= \lambda_k e - \alpha \eta_{k-1},\quad \alpha \geq 0 \\ z_k &= \text{argmin}_z\,\{\|z - y_k\| | z \in M\} \\ r_k &= \|z_k - y_k\| \\ \eta_k &= (z_k - y_k)/r_k \\ \lambda_{k+1} &= (\eta_k, z_k)/(\eta_k, e) \end{aligned} \tag{3.44}$$

From the fact that z_k is a minimum of $\|z-y_k\|$ on M it follows that the equation

$$(\eta_k, z) \geq (\eta_k, z_k) = \lambda_{k+1}(\eta_k, e), \quad z \in M \tag{3.45}$$

is satisfied, so that λ_{k+1}, η_k and z_k satisfy an equation analogous to (3.43).

We note to begin with, that by virtue of (3.43), $(\eta_{k-1}, z_k - \lambda_k e) \geq 0$; thus

$$\begin{aligned} r_k^2 &= \|z_k - y_k\|^2 \\ &= \|z_k - \lambda_k e\|^2 + 2(z_k - \lambda_k e, \lambda_k e - y_k) + \|\lambda_k e - y_k\|^2 \\ &= \|z_k - \lambda_k e\|^2 + 2\alpha(z_k - \lambda_k e, \eta_{k-1}) + \alpha^2 \geq \alpha^2. \end{aligned}$$

Thus, we always have $r_k \geq \alpha$. Here, if $r_k = \alpha$, then $\|z_k - \lambda_k e\| = 0$ and $z_k = \lambda_k e \in M$ and since $\lambda_k \leq \lambda_*$, $\lambda_k = \lambda_*$ is a solution of problem (3.42). Conversely, if $\lambda_k = \lambda_*$, then $\lambda_k e \in M$ and, by virtue of the choice of z_k

$$\begin{aligned} r_k^2 &= \|z_k - \lambda_k e\|^2 + 2\alpha(z_k - \lambda_k e, \eta_{k-1}) + \alpha^2 \\ &\leq \|\lambda_k e - y_k\|^2 = \alpha^2 \end{aligned}$$

i.e. $r_k \leq \alpha$.

Thus, the equation $r_k = \alpha$ is a sign of the fact that problem (3.42) is solved.

In order to justify the operation of the algorithm we have to show that if $\lambda_k < \lambda_*$ then $\lambda_{k+1} \leq \lambda_*$ and that (η_k, e) is strictly positive.

Suppose that $\lambda_k < \lambda_*$. It follows from (3.43) that

$$(\eta_{k-1}, z_k) \geq (\eta_{k-1}, \lambda_k e) = (\eta_{k-1}, y_k) + \alpha(\eta_{k-1}, \eta_{k-1})$$

or, since $\|\eta_{k-1}\| = 1$,

$$(\eta_{k-1}, z_k - y_k) = r_k(\eta_{k-1}, \eta_k) \geq \alpha. \tag{3.46}$$

Thus, it is clear that $(\eta_{k-1}, \eta_k) \geq 0$ and $r_k \geq \alpha$. Moreover, it follows from (3.45) that

$$(\eta_k, z) \geq (\eta_k, z_k - y_k) + (\eta_k, y_k) = r_k + \lambda_k(\eta_k, e) - \alpha(\eta_k, \eta_{k-1}).$$

Substituting $z = \lambda_* e$, we obtain

$$(\lambda_* - \lambda_k)(\eta_k, e) \geq r_k - \alpha(\eta_k, \eta_{k-1}) \geq r_k - \alpha \geq 0. \tag{3.47}$$

Thus,

$$(\eta_k, e) > 0$$

since $(\eta_k, e) = 0$ would imply that $r_k = \alpha$ and $\lambda_k = \lambda_*$. Then, substituting $z = \lambda_* e$ in (3.45), we obtain

$$(\lambda_* - \lambda_{k+1})(\eta_k, e) \geq 0$$

i.e. $\lambda_{k+1} \leq \lambda_*$.

Let us formalize the result obtained above.

Theorem 3.7 *If $\alpha \geq 0$ and $\lambda_0 < \lambda_*$ then the algorithm (3.44) for $k = 1, 2, \ldots$ generates a sequence λ_k such that $\lambda_k \leq \lambda_*$ and $r_k \geq \alpha$. Here, satisfaction of the condition $r_k = \alpha$ at some step of the algorithm implies that $\lambda_k = \lambda_*$ and $z_k = \lambda_* e$, i.e. the condition $r_k = \alpha$ is a sign that the algorithm has stopped.*

From formula (3.44) it follows that

$$\begin{aligned}\lambda_{k+1} &= \frac{(\eta_k, z_k - y_k) + \lambda_k(\eta_k, e) - \alpha(\eta_k, \eta_{k-1})}{(\eta_k, e)} \\ \lambda_{k+1} &= \lambda_k + \frac{r_k - \alpha(\eta_k, \eta_{k-1})}{(\eta_k, e)} \end{aligned} \tag{3.48}$$

Since $\|\eta_k\| = \|\eta_{k-1}\| = 1$, we have

$$\|\eta_k - \eta_{k-1}\|^2 = 2(1 - (\eta_k, \eta_{k-1})).$$

Thus, (3.48) may be rewritten in the form

$$\lambda_{k+1} = \lambda_k + \frac{r_k - \alpha}{(\eta_k, e)} + \alpha \frac{\|\eta_k - \eta_{k-1}\|^2}{2(\eta_k, e)} \tag{3.49}$$

Whence it follows that λ_k is an increasing sequence.

Theorem 3.8 *The sequence λ_k generated by the algorithm is monotonic increasing and converges to λ_*. Here, $z_k \to \lambda_* e$ and any limit point η of the sequence η_k is a support vector for M at the point $\lambda_* e$, i.e.*

$$(z - \lambda_* e, \eta) \geq 0, \quad z \in M.$$

Proof. As previously shown, if the algorithm stops for some k (i.e. $r_k = \alpha$) then $\lambda_k = \lambda_*$ and $z_k = \lambda_* e$. Thus, we shall consider the case where $r_k > \alpha$ for all k and the process is infinite.

From (3.49), we deduce that $\lambda_{k+1} - \lambda_k \geq r_k - \alpha$, since $0 < (\eta_k, e) \leq 1$. From the fact that the sequence λ_k is monotonic increasing and bounded above, it follows that $\lambda_{k+1} - \lambda_k \to 0$, whence $r_k \to \alpha$. As previously shown,

$$\begin{aligned} r_k^2 &= \|z_k - \lambda_k e\|^2 + 2\alpha(z_k - \lambda_k e, \eta_{k-1}) + \alpha^2 \\ (z_k - \lambda_k e, \eta_{k-1}) &\geq 0 \end{aligned}$$

i.e. $r_k^2 - \alpha^2 \geq \|z_k - \lambda_k e\|^2$, whence $\|z_k - \lambda_k e\| \to 0$.

If $\underline{\lambda}$ denotes the limit of the sequence λ_k (which is monotonic increasing and bounded above), then $z_k \to \underline{\lambda} e$, $\underline{\lambda} \leq \lambda_*$. But $z_k \in M$ and M is closed. Thus, $\underline{\lambda} e \in M$, since $\underline{\lambda} = \lambda_*$.

The last assertion of the theorem is obtained immediately by passing to the limit in (3.45). □

Remark. If M is not closed then $\underline{\lambda} e \in \overline{M}$ where $\overline{M}$ denotes the closure of M and $\underline{\lambda} = \min\{\lambda | \lambda e \in \overline{M}\}$.

It is not difficult to see this.

Theorem 3.9 *If M is a polyhedral set then the algorithm converges in a finite number of steps.*

Proof. Suppose that M is defined by a finite system of linear inequalities

$$(a_i, z) - b_i \leq 0, \quad i \in I. \tag{3.50}$$

For $z \in M$, we denote $I(z) = \{i \in I | (a_i, z) = b_i\}$. As previously shown, $z_k \to \lambda_* e$. Thus, $I(z_k) \subseteq I(\lambda_* e)$ for k sufficiently large. On the other hand, z_k is a minimum of the function $1/2\|z - y_k\|^2$ on the set M defined by the inequalities (3.50); thus by virtue of the necessary conditions for a minimum

$$z_k - y_k = - \sum_{i \in I(z_k)} \lambda_i a_i, \quad \lambda_i \geq 0$$

or

$$\eta_k = - \sum_{i \in I(z_k)} \gamma_i a_i, \quad \gamma_i = \lambda_i / r_k \geq 0.$$

Since $I(z_k) \subseteq I(\lambda_* e)$, then

$$(a_i, z_k) = (a_i, \lambda_* e) = b_i, \quad i \in I(z_k)$$

whence

$$(\eta_k, \lambda_* e) = (\eta_k, z_k)$$

in other words

$$\lambda_* = (\eta_k, z_k)/(\eta_k, e) = \lambda_{k+1}$$

and the finite convergence of the algorithm now follows. □

3.2.2 Bounds on the Rate of Convergence

As usual, in order to give more precise bounds on the rate of convergence we must make a number of additional assumptions.

Since the sets $\{\lambda e | \lambda < \lambda_*\}$ and M do not intersect, they may be separated; in other words, there exists a vector e_1 such that

$$(\lambda e, e_1) \leq (z, e_1), \quad \lambda < \lambda_*, \quad z \in M.$$

Thus, it follows that (e, e_1) is nonnegative and that

$$(\lambda_* e, e_1) \leq (z, e_1), \quad z \in M. \tag{3.51}$$

We shall assume that there exists a vector e_1 satisfying (3.51) and the strict inequality

$$(e, e_1) > 0. \tag{3.52}$$

We substitute the vector

$$z_k = y_k + r_k\eta_k = \lambda_k e + r_k\eta_k - \alpha\eta_{k-1}$$

into (3.51) instead of z. We obtain

$$r_k(e_1, \eta_k) \geq (\lambda_* - \lambda_k)(e, e_1) + \alpha(e_1, \eta_{k-1}) \tag{3.53}$$

or, more coarsely,

$$\begin{aligned} r_k &\geq \delta_k(e, e_1) + \alpha(e_1, \eta_{k-1}) \\ \delta_k &= \lambda_* - \lambda_k. \end{aligned} \tag{3.54}$$

Discarding the last term in the expression (3.49) for λ_{k+1} and replacing r_k by (3.54), we obtain

$$\lambda_{k+1} \geq \lambda_k + \delta_k \frac{(e, e_1)}{(e, \eta_k)} - \alpha \frac{1 - (e_1, \eta_{k-1})}{(e, \eta_k)}$$

or

$$\delta_{k+1} \leq \left(1 - \frac{(e, e_1)}{(e, \eta_k)}\right) \delta_k + \alpha \frac{\|e_1 - \eta_{k-1}\|^2}{2(e, \eta_k)}. \tag{3.55}$$

Formula (3.55) enables us to draw a number of immediate conclusions.

Theorem 3.10 *Suppose that $\alpha = 0$ and that the vector e_1 satisfying equation (3.51) is unique. Suppose also that the inequality (3.52) is satisfied. Then δ_k converges to zero superlinearly.*

Proof. In fact, under the conditions of Theorem 3.10, it follows from Theorem 3.8 that $\eta_k \to e_1$. Thus, $\delta_{k+1} \leq \gamma_k\delta_k$ and $\gamma_k = 1 - (e, e_1)/(e, \eta_k) \to 0$. □

We shall study the case in which $\alpha > 0$ in more detail. Let us assume that the set M is sufficiently smooth in a neighbourhood of the point $\lambda_* e$, i.e. it may be defined in a neighbourhood of this point by an inequality $\varphi(z) \leq 0$, where φ is a twice continuously differentiable convex function. Clearly, in this case

$$e_1 = -\varphi'(\lambda_* e)\|\varphi'(\lambda_* e)\|^{-1}$$

where $\varphi'(z)$ is the gradient of the function φ.

We shall then represent each vector z in the form

$$z = \beta e + w, \quad (e, w) = 0.$$

It is easy to see that

$$\beta = (z, e), \quad w = z - (z, e)e.$$

We denote the projection matrix for projection onto the subspace $(e, z) = 0$ by $P = I - ee^*$, where e^* is the transpose of e (i.e. a row vector). Clearly,

$$z = (I - P)z + Pz, \quad \beta e = (I - P)z, \quad w = Pz.$$

If condition (3.52) is satisfied then it is easy to show that, in a neighbourhood of the point $\lambda_* e$, the set M is described by the inequality

$$\beta \geq \omega(w) \tag{3.56}$$

where ω is a twice continuously differentiable function. Here, since $\beta = \lambda_*$ and $w = 0$ for $z = \lambda_* e$, we have

$$\lambda_* = \omega(0).$$

Suppose that z is a boundary point of M, i.e. $\beta = \omega(w)$. We denote the unit normal to M at z by $\eta(z)$. It is easy to see that

$$\eta(z) = \frac{1}{\sqrt{1 + \|\omega'(w)\|^2}} \begin{pmatrix} 1 \\ -\omega'(w) \end{pmatrix}$$

where $\omega'(w)$ is the derivative of ω with respect to w, where the dimension of w is one less than the dimension of z. Moreover, $\eta(0) = e_1$. Let us find the point of intersection of the tangent hyperplane to M at the point z and the line λe. In the coordinates (β, w), the corresponding equation has the form

$$\begin{aligned} -(\lambda - \beta) + (\omega'(w), -w) &= 0 \\ \lambda(z) &= \beta - (\omega'(w), w) = \omega(w) - (\omega'(w), w). \end{aligned}$$

Thus, decomposition of $\omega(w)$ and $\omega'(w)$ into Taylor series in w in a neighbourhood of the point $w = 0$ up to second order terms gives

$$\begin{aligned} \lambda(z) &= \omega(0) - (\omega''(0)w, w)/2 + \mathrm{o}\left(\|w\|^2\right) \\ \delta(z) &= \lambda_* - \lambda(z) = (\omega''(0)w, w)/2 + \mathrm{o}\left(\|w\|^2\right). \end{aligned}$$

Suppose now that

$$\mu = \min_{w \neq 0} \frac{(\omega''(0)w, w)}{\|w\|^2} > 0.$$

Then

$$\delta(z) \geq (\mu/4)\|w\|^2 \tag{3.57}$$

for any point z on the boundary of M, for w sufficiently small. Moreover, if w is given, then the boundary point z of the set M corresponding to it has coordinates $\beta = \omega(w)$. Thus, the normal $\eta(z)$ defined at the boundary point z is a function w and, since $\omega(w)$ is twice continuously differentiable, it follows that ω as a function of w satisfies a Lipschitz condition with constant L in a neighbourhood of $w = 0$.

By virtue of (3.57), we obtain

$$\frac{\|e_1 - \eta(z)\|^2}{2} = \frac{\|\eta(0) - \eta(z)\|^2}{2} \le \frac{L^2}{2}\|w\|^2 \le \frac{2L^2}{\mu}\delta(z). \tag{3.58}$$

We now use this bound to transform the bound (3.57). Since $\eta_{k-1} = \eta(z_{k-1})$ and $\delta_{k-1} = \delta(z_{k-1})$, then

$$\delta_{k+1} \le \left(1 - \frac{(e, e_1)}{(e, \eta_k)}\right)\delta_k + \frac{2L^2}{\mu}\alpha\frac{\delta_{k-1}}{(e, \eta_k)}. \tag{3.59}$$

Theorem 3.11 *Suppose that in a neighbourhood of the point $\lambda_* e$, the set M may be defined by the inequality (3.56), where $\omega(w)$ is a twice continuously differentiable convex function and*

$$\mu = \min_{\substack{w \neq 0 \\ (e,w)=0}} \frac{(\omega''(0)w, w)}{\|w\|^2} > 0.$$

Then the rate of convergence of the given algorithm for large k is determined by formula (3.59).

Since the sequence δ_k is decreasing and since under the given assumptions $\eta_k \to \eta(0) = e_1$, it follows from (3.59) that

$$\limsup_{k \to \infty} \frac{\delta_{k+1}}{\delta_{k-1}} \le \frac{2L^2}{\mu(e, e_1)}\alpha. \tag{3.60}$$

Whence it is clear that decreasing α leads to an acceleration of the process of convergence.

We shall show that a smaller constant L may be obtained by a simple transformation, which also leads to an acceleration of the convergence. We carry out a coordinate transformation which reduces to extending the component of the vector z orthogonal to the vector e by a factor K. If P denotes the operator of orthogonal projection onto the hyperplane orthogonal to e then

$$P = I - ee*, \quad P^2 = P, \quad P(I - P) = 0.$$

In what follows, we shall denote all values in the new coordinates by a tilde. The coordinate transformation is then given by the formula

$$\tilde{z} = (I - P)z + KPz$$

and it is easy to see that

$$z = (I - P)\tilde{z} + K^{-1}P\tilde{z} = (ee^* + K^{-1}P)\tilde{z}.$$

Moreover

$$\tilde{w} = P\tilde{z} = KPz = Kw.$$

Let us now consider how a vector normal to the set M at a boundary point of this set is transformed. We recall that the boundary point z was uniquely defined by its coordinates $w = Pz$, since the coordinate β along the e axis is found from the equation $\beta = \omega(w)$.

Suppose that η denotes the unit normal to M at the point z, i.e.

$$(\eta, z_1 - z) \geq 0, \quad z_1 \in M, \quad \|\eta\| = 1.$$

Substituting

$$z_1 = (ee^* + K^{-1}P)\tilde{z}_1, \quad z = (ee^* + K^{-1}P)\tilde{z}$$

we obtain

$$(\eta, (ee^* + K^{-1}P)(\tilde{z}_1 - \tilde{z})) \geq 0, \quad \tilde{z}_1 \in \widetilde{M} \tag{3.61}$$

where $\widetilde{M}$ denotes the set M in the new coordinates. Since P is a symmetric matrix, (3.61) may be rewritten in the form

$$(\tilde{\eta}, \tilde{z}_1 - \tilde{z}) \geq 0, \quad \tilde{z}_1 \in \widetilde{M} \tag{3.62}$$

where

$$\tilde{\eta} = \frac{(ee^* + K^{-1}P)\eta}{\|(ee^* + K^{-1}P)\eta\|}.$$

Since the vectors e and $P\eta$ are orthogonal, it is easy to show that

$$\|(ee^* + K^{-1}P)\eta\| = \sqrt{(e, \eta)^2 + K^{-2}(1 - (e, \eta)^2)}.$$

Using this together with (3.62), we finally obtain that the vector $\tilde{\eta}(\tilde{w})$ in the new coordinates is expressed in terms of the vector $\eta(w)$ by the formula

$$\tilde{\eta}(\tilde{w}) = \frac{(e, \eta(w))e + K^{-1}P\eta(w)}{\sqrt{(e, \eta(w))^2 + K^{-2}(1 - (e, \eta(w))^2)}}. \tag{3.63}$$

Since $\tilde{e} = e$, $e_1 = \eta(0)$ and $(Pe_1, e) = 0$, it follows from (3.63) that for $K > 1$

$$\begin{aligned}
(\tilde{e}, \tilde{e}_1) &= \frac{(e, e_1) + K^{-1}(Pe_1, e)}{\sqrt{(e, e_1)^2 + K^{-2}(1 - (e, e_1)^2)}} \\
&= \frac{(e, e_1)}{\sqrt{(e, e_1)^2 + K^{-2}(1 - (e, e_1)^2)}} \\
&> (e, e_1)
\end{aligned}$$

i.e.

$$(\tilde{e}, \tilde{e}_1) > (e, e_1). \tag{3.63'}$$

Thus, the scalar product on the right-hand side of (3.63′) increases under the coordinate transformation.

On the other hand, relatively uncomplicated, but tedious analysis of expression (3.63) shows that in the new coordinates the Lipschitz constant $\widetilde{L}$ for $\tilde{\rho}(\tilde{w})$ is expressed in terms of L by the formula

$$\widetilde{L} = CL/K^2$$

for some number C. Moreover, since after the transformation $\tilde{\beta} = \beta$ and $\tilde{w} = Kw$, the inequality

$$\beta \geq \omega(w)$$

which defines M in a neighbourhood of the point $\lambda_* e$ becomes the inequality

$$\tilde{\beta} \geq \omega(\tilde{w}K) = \tilde{\omega}(\tilde{w})$$

we have

$$\tilde{\omega}''(0) = K^{-2}\omega''(0).$$

Thus,

$$\tilde{\mu} = K^{-2}\mu.$$

Therefore, in the new coordinates, formula (3.60) may be rewritten in the form

$$\begin{aligned}\limsup_{k\to\infty}\frac{\tilde{\delta}_{k+1}}{\delta_{k-1}} &\leq \frac{2(\widetilde{L})^2}{\tilde{\mu}(\tilde{e},\tilde{e}_1)}\alpha \\ &\leq \frac{2C^2L^2}{\mu(e,e_1)}K^{-2}\alpha. \qquad (3.64)\end{aligned}$$

It is clear from this formula that extension of the coordinates in the hyperplane orthogonal to e leads to a considerable acceleration of the convergence for large K.

3.2.3 An Algorithm for Convex Programming Problems

Let us now return to the original problem

$$\min_x\{f_0(x)|f_i(x) \leq 0,\ i = 1,\dots,m,\ x \in X\} \qquad (3.65)$$

and consider how the given algorithm reduces. As previously mentioned at the start of the section, for this problem

$$M = \{f(x) + v|x \in X, v \geq 0\}$$

where $f(x) \in \mathbb{R}^{m+1}$ is the vector with components $f_i(x)$, $i = 0, 1, \dots, m$ and v is the vector with components v_i, $i = 0, \dots, m$. It is easy to show that if the $f_i(x)$ are continuous convex functions, the set X is convex and closed and the set

$$X_e = \{x \in X | f_0(x) \leq C,\ f_i(x) \leq 0,\ i = 1, \dots, m\}$$

is bounded for some C, then M is a closed convex set and problem (3.65) has at least one solution x_*. Here, $\lambda_* = f_0(x_*)$,

$$e = \begin{pmatrix} 1 \\ 0 \\ \dots \\ 0 \end{pmatrix} \in \mathbb{R}^{m+1}, \quad z_* = \lambda_* e = \begin{pmatrix} f_0(x_*) \\ 0 \\ \dots \\ 0 \end{pmatrix} \in \mathbb{R}^{m+1}$$

Suppose that there exists a Kuhn–Tucker vector $u \in \mathbb{R}^m$, i.e.

$$\begin{aligned} u \geq 0,\ u^i f_i(x_*) &= 0, \quad i = 1, \dots, m \\ f_0(x_*) + \sum_{i=1}^{m} u^i f_i(x_*) &\leq f_0(x) + \sum_{i=1}^{m} u^i f_i(x), \quad x \in X. \end{aligned}$$

It is easy to see that, in this case, the vector

$$e_1 = \begin{pmatrix} 1/\sqrt{1 + \|u\|^2} \\ u^1/\sqrt{1 + \|u\|^2} \\ \dots \\ u^m/\sqrt{1 + \|u\|^2} \end{pmatrix} \in \mathbb{R}^{m+1}$$

is the unit normal to M at the point z_*.

The basic procedure described in stage 1 of the algorithm amounts to finding a point $z_k \in M$ near to the point y_k. By virtue of the particular structure of the set M, we obtain

$$\begin{aligned} r_k^2 &= \min_{z \in M} \sum_{i=0}^{m} (z^i - y_k^i)^2 \\ &= \min_{\substack{x \in X \\ v \geq 0}} \sum_{i=0}^{m} (f_i(x) + v^i - y_k^i)^2 \\ &= \min_{x \in X} \sum_{i=0}^{m} (f_i(x) - y_k^i)_+^2 \end{aligned} \tag{3.66}$$

where, as usual, $t_+ = \max(0, t)$. If the minimum of (3.66) is attained at the point x_k, then the corresponding point z_k has the form

$$z_k^j = \begin{cases} y_k^i & \text{if } y_k^i - f_i(x_k) \geq 0 \\ f_i(x_k) & \text{if } y_k^i - f_i(x_k) < 0 \end{cases}. \tag{3.67}$$

Then it is easy to calculate the vector $\eta_k = r_k^{-1}(z_k - y_k)$:

$$\eta_k^i = r_k^{-1}(f_i(x_k) - y_k^i)_+, \quad i = 0, \ldots, m. \tag{3.68}$$

Thus, for problem (3.65), the algorithm takes on the following form: at the step with index $k \geq 0$

$$\begin{aligned}
y_k &= \lambda_k e - \alpha \eta_{k-1}, \quad k, \alpha = 0 \\
x_k &= \operatorname{argmin} \{\sum_{i=0}^{m} (f_i(x) - y_k^i)_+^2 | x \in X\} \\
\eta_k^i &= r_k^{-1}(f_i(x_k) - y_k^i)_+, \quad i = 0, \ldots, m \\
\lambda_{k+1} &= \frac{1}{\eta_k^0} \sum_{i=0}^{m} \eta_k^i z_k^i \\
&= \frac{1}{\eta_k^0} \sum_{i=0}^{m} [f_i(x_k) + (y_k^i - f_i(x_k))_+] \eta_k^i \\
&= \frac{1}{\eta_k^0} \sum_{i=0}^{m} f_i(x_k) \eta_k^i
\end{aligned} \tag{3.69}$$

where (3.67) and (3.68) were used to derive the formula for λ_{k+1}. The value of λ_0 should be chosen using the condition $\lambda_0 \leq f_0(x_*)$. The stopping criterion is the satisfaction of the inequality $r_k - \alpha \leq \epsilon$, where $\epsilon > 0$ is a previously assigned accuracy.

It follows from the previous theorem that the algorithm always converges. If the original problem is a linear programming problem it converges in a finite number of steps.

Let us now consider the question of the convergence in the general case. The subset M of $\mathbb{R}^{m+1}$ may be considered as the epigraph of some function $\omega(w)$, where $w \in \mathbb{R}^n$ is a vector with components $z^1, \ldots, z^m$. Then

$$\begin{aligned}
\omega(w) &= \min_z \{z^0 | (z^0, z^1, \ldots, z^m)^* \in M\} \\
&= \min\{f_0(x) | f_i(x) \leq z^i, \ i = 1, \ldots, m, \ x \in X\} \\
M &= \{z | z^0 \geq \omega(w)\}.
\end{aligned}$$

It is known that $\omega(w)$ is a convex function. Here, $\lambda_* = f_0(x_*) = \omega(0)$.

Theorem 3.12 *If the function $\omega(w)$ is twice continuously differentiable at the point $w = 0$ and*

$$(\omega''(0)w, w) \geq \mu \|w\|^2, \quad \mu > 0$$

then the rate of convergence of the algorithm (3.69) is determined by the equation

$$\limsup_{k \to \infty} \frac{\delta_{k+1}}{\delta_{k-1}} \leq \frac{2L^2}{\mu} \sqrt{1 + \|u\|^2} \alpha$$

where L is the Lipschitz constant for the unit normal to M in a neighbourhood of the point $z^0 = f_0(x_*)$, $z^i = 0$, $i = 1, \ldots, m$, and $u \in \mathbb{R}^m$ is the unique vector of the Kuhn–Tucker problem (3.65).

Proof. Since the function $\omega(w)$ is smooth, it follows from the known results of Section 1.2 that the Kuhn–Tucker vector is uniquely determined and is given by the vector $-\omega'(0)$. Moreover, by virtue of the particular form of the vectors e and e_1 in the present case,

$$(e, e_1) = 1/\sqrt{1 + \|u\|^2}.$$

Theorem 3.12 is then obtained as a direct consequence of Theorem 3.11 and formula (3.60). □

This result characterizes the convergence to zero of δ_k. In practice, we are very interested in the convergence of the points x_k to the minimum x_* or, at least, in the behaviour of the values $f_i(x_k)$, $i = 0, 1, \ldots, m$.

We recall that $\delta_k = \lambda_* - \lambda_k = f_0(x_*) - \lambda_k$. On the other hand, it follows from formula (3.57) that

$$\sum_{i=1}^{m} \left(z_k^i\right)^2 \leq 4\delta_k/\mu$$

and, in particular,

$$|z_k^i| \leq \sqrt{4\delta_k/\mu}.$$

But formula (3.67) shows that $f_i(x_k) \leq |z_k^i|$. Thus,

$$f_i(x_k) \leq \sqrt{4\delta_k/\mu}, \quad i = 1, \ldots, m$$

so that the extent to which the point x_k satisfies the constraints depends on the value of $\sqrt{\delta_k}$.

As Theorem 3.12 shows, the rate of convergence of the algorithm is determined by the value of α; the smaller $\alpha > 0$, the greater the convergence.

In Section 3.2.2 we described yet another technique for accelerating the convergence by expansion in the subspace orthogonal to the vector e. Because of the specific form of this vector, it is easy to see that this expansion in this case reduces to multiplying all the coordinates z^i, $i = 1, \ldots, m$ by the value $K \geq 1$, i.e. in the case of problem (3.65), to replacing all the functions $f_i(x)$, $i = 1, \ldots, m$ by functions $Kf_i(x)$. It is also easy to show that, in this case, the Kuhn–Tucker vector u is replaced by the vector $K^{-1}u$. Using the results of Section 3.2.2 and, in particular, formula (3.64), we see that the following theorem holds.

Theorem 3.13 *Suppose that the conditions of Theorem 3.12 are satisfied. Then the algorithm described by the formulae of (3.69), with all the functions $f_i(x)$, $i = 1, \ldots, m$, replaced by functions $Kf_i(x)$, $K \geq 1$ converges at a rate determined by*

$$\limsup_{k \to \infty} \frac{\delta_{k+1}}{\delta_{k-1}} \leq \frac{2C^2L^2}{\mu} \sqrt{1 + K^2\|u\|^2} \frac{\alpha}{K^2}.$$

We shall now state a theorem which guarantees that the conditions of Theorem 3.12 are satisfied.

Theorem 3.14 *Suppose that x_* is a solution of problem (3.65) and the following conditions are satisfied:*

a) the functions $f_i(x)$ are convex and twice continuously differentiable;

b) the gradients $f_i'(x)$, $i \in I_0$, where

$$I_0 = \{i | f_i(x_*) = 0, \ i = 1, \dots .m\}$$

are linearly independent and the components of the Kuhn–Tucker vector are strictly positive;

c) the matrix of second derivatives with respect to x of the Lagrange function

$$L(x, u) = f_0(x) + \sum_{i=1}^{m} u^i f_i(x)$$

is strictly positive definite at the point x_*, i.e.

$$(L''_{xx}(x_*, u)p, p) > 0, \quad p \neq 0.$$

Then the conditions of Theorem 3.12 are satisfied.

Proof. Without loss of generality, we shall assume that the set I_0 contains all the indices $i = 1, \dots .m$. For w in a neighbourhood of zero, we define a system of equations

$$L'_x(x(w), u(w)) = 0, \quad \tilde{f}(x(w)) - w = 0 \tag{3.70}$$

where w is the vector with components $z^1, \dots, z^m$ and $\tilde{f}(x)$ has components $f_1(x), \dots, f_m(x)$. In this system, x and u are considered as unknown functions of the vector w. For $w = 0$, the system (3.70) is a list of the necessary and sufficient conditions for an extremum in problem (3.65), and thus, by the assumptions of the theorem, has a unique solution $x(0) = x_*$, $u(0) = u > 0$. On the other hand, the matrix of the derivatives of the left-hand side of the system (3.70) with respect to x and u at the point $x = x(0)$, $u = u(0)$ has the form

$$\begin{pmatrix} L''_{xx}(x_*, u_*) & \tilde{f}'^*(x_*) \\ \tilde{f}'(x_*) & 0 \end{pmatrix}$$

where $\tilde{f}'(x_*)$ is an $m \times n$ matrix with components $\partial f_i(x)/\partial x^j$, $i = 1, \dots, m$, $j = 1, \dots, n$ where x^j denotes the jth component of the vector $x \in \mathbb{R}^n$. As shown in Section 2.3, under the given assumptions, this matrix is nonsingular; thus, by the implicit function theorem, the system (3.70) has a unique solution in a neighbourhood of $w = 0$. Here, the matrices $x'_w(0)$ and $u'_w(0)$ of the derivatives of x and u with respect to w have the following form

$$x'_w(0) = \begin{pmatrix} \frac{\partial x^1(0)}{\partial z^1} & \dots & \frac{\partial x^1(0)}{\partial z^m} \\ \dots & \dots & \dots \\ \frac{\partial x^n(0)}{\partial z^1} & \dots & \frac{\partial x^n(0)}{\partial z^m} \end{pmatrix}$$

$$u'_w(0) = \begin{pmatrix} \frac{\partial u^1(0)}{\partial z^1} & \dots & \frac{\partial u^1(0)}{\partial z^m} \\ \dots & \dots & \dots \\ \frac{\partial u^m(0)}{\partial z^1} & \dots & \frac{\partial u^m(0)}{\partial z^m} \end{pmatrix}$$

and satisfy the equations

$$\begin{aligned} L''_{xx}(x_*, u)x'_w(0) + \tilde{f}'^*(x_*)u'_w(0) &= 0 \\ \tilde{f}'(x_*)x'_w(0) - I_m &= 0 \end{aligned}$$

where I_m is the unit $m \times m$ matrix. Thus, we obtain

$$u'_w(0) = -\left(\tilde{f}'(x_*)(L''_{xx}(x_*, u))^{-1}\tilde{f}'^*(x_*)\right)^{-1}$$

Under the assumptions of Theorem 3.14, the matrix $u'_w(0)$ is strictly negative definite.

Since $u = u(0) > 0$ it follows that $u(w) > 0$ for small w and $x(w)$ is a minimum of $f_0(x)$ subject to the constraints $f_i(x) \le z^i$, $i = 1, \dots, m$ with corresponding Kuhn–Tucker vector $u(w)$.

As previously mentioned, $u(w) = -\omega'(w)$. Thus, $\omega''(0) = -u'_w(0)$ and so the matrix $\omega''(0)$ is strictly positive definite; this proves that the conditions of Theorem 8.6 are satisfied. □

As far as successful operation of this algorithm is concerned, the quality and effectiveness of the solution of the auxiliary problem are important. As in the method of modified Lagrange functions, two forms of difficulty arise here.

In (3.66), the function to be minimized has discontinuous second order derivatives. Thus, it is desirable that, at least in a neighbourhood of the solution, the function should not have such discontinuities. The second difficulty is associated with the extent to which this function is singular.

Let us consider how the function

$$\sum_{i=0}^{m} (f_i(x) - y_k^i)_+^2$$

behaves for k sufficiently large, i.e. $y_k \approx \lambda_* e - \alpha e_1$. Then

$$\begin{aligned} y_k^0 &\approx f_0(x_*) - \alpha/\sqrt{1 + \|u\|^2} \\ y_k^i &\approx -\alpha u^i/\sqrt{1 + \|u\|^2}, \quad i = 1, \dots, m. \end{aligned}$$

Thus, for large k, the auxiliary problem reduces to minimization of a function of type

$$\left(f_0(x) - f_0(x_*) + \frac{\alpha}{\sqrt{1 + \|u\|^2}}\right)_+^2 + \sum_{i=1}^{m} \left(f_i(x) + \frac{\alpha u^i}{\sqrt{1 + \|u\|^2}}\right)_+^2.$$

If $\alpha > 0$ and $u^i > 0$ for $i \in I_0$, then for x in some neighbourhood of x_*, this last function takes the form

$$\left(f_0(x) - f_0(x_*) + \frac{\alpha}{\sqrt{1+\|u\|^2}}\right)^2 + \sum_{i\in I_0}\left(f_i(x) + \frac{\alpha u^i}{\sqrt{1+\|u\|^2}}\right)^2$$

i.e. it is a smooth function. Let us compute its matrix of second derivatives at the point x_*. A simple calculation shows that this matrix is equal to

$$A = \frac{2\alpha}{\sqrt{1+\|u\|^2}} L''_{xx}(x_*, u) + 2\sum_{i\in I_0} f'_i(x_*)(f'_i(x_*))^* \tag{3.71}$$

where $f'_i(x_*)$ is a column vector and $(f'_i(x_*))^*$ is a row vector.

The given algorithm represents the external cycle of the solution of the problem. It includes the internal cycle which involves the minimization problem (3.66). If the set X has a simple structure, in particular, if $X = \mathbb{R}^n$ then good, tried and tested, standard algorithms may be used to solve problem (3.66).

By virtue of the fact that the matrix L''_{xx} is strictly positive definite and from expression (3.71), it follows that the matrix A is strictly positive definite. This is an important condition for high-speed convergence for all unconstrained-optimization algorithms. Here, that A is strictly positive definite is a result of the fact that $\alpha > 0$. From what has gone before, it is clear that the requirements on the quality of convergence of the external and internal cycles lead to contradictory requirements on α. For fast convergence of the external cycle, α should decrease to zero, while for fast convergence of the internal cycle, α should be sufficiently large. We recall that it is also possible to accelerate the convergence by multiplying all the functions $f_i(x)$, $i = 1, \ldots, m$ by $K > 0$. It is evidently impossible to give general recommendations for the choice of $\alpha > 0$ and $K > 1$. However, judicious variation of these values on the whole ensures that the process converges rapidly.

3.3 Algorithms and Examples

3.3.1 The Linearization Method

In Section 2.1 we mentioned a number of computational aspects of the linearization method. We shall now make additional recommendations relating to the practical side of the question and give examples of the calculations.

When describing the linearization algorithm for a practical implementation, it is convenient to formulate the minimization problem (2.1) in the following form

It is desired to minimize $f_0(x)$, $x \in E^n$, subject to the constraints

$$\begin{aligned} f_i(x) &\leq 0, \quad i = 1, \ldots, m \\ f_i(x) &= 0, \quad i = m+1, \ldots, m+l. \end{aligned} \tag{3.72}$$

We denote

$$\begin{aligned} F(x) &= \max\{0, f_1(x), \ldots, f_m(x), |f_{m+1}(x)|, \ldots, |f_{m+l}(x)|\} \\ I_\delta^-(x) &= \{i | f_i(x) \geq F(x) - \delta,\ i = 1, \ldots, m\} \\ I_\delta^0(x) &= \{i || f_i(x)| \geq F(x) - \delta,\ i = m+1, \ldots, m+l\} \\ \Phi_N(x) &= f_0(x) + NF(x),\ \ \|p\|^2 = \sum_{i=1}^{n} (p_i)^2. \end{aligned}$$

The linearization method generates two iterative sequences, namely $\{x_k\}$, $k = 1, 2, \ldots$ which converges to a solution of problem (3.72) x_* and $\{u_k\}$, $k = 1, 2, \ldots$ which converges to a Lagrange multiplier u_*. The sequence $\{u_k\}$ is generated by solving the dual problem to problem (2.4).

Let us now write this problem in a form suitable for implementation on a computer. We expand the norm in expression (2.21) and bring it to the form

$$\varphi(u) = -1/2(Au, u) + (d, u) + C \tag{3.73}$$

Here, the matrix A is symmetric with elements

$$\{a_{ij}\} = \{f_i'(x) f_j'^*(x)\},\ \ i, j \in I_\delta^-(x) \cup I_\delta^0(x),$$

the ith component of the vector d is equal to $f_0'(x) f_i'^*(x) + f_i(x)$, $i \in I_\delta^-(x) \cup I_\delta^0(x)$ and $C = 1/2 f_0'(x) f_0'^*(x)$. We obtain the quadratic programming problem

$$\max_u \{\varphi(u) | u^i \geq 0,\ i \in I_\delta^-(x)\}. \tag{3.74}$$

This problem may be solved by the conjugate gradient method described in Section 1.3.8 or by any other finite method for solving quadratic programming problems.

We shall now describe the linearization algorithm. Suppose that an arbitrary initial approximation x_0, an accuracy for solution of the original problem (3.72) ϵ_1, an accuracy for solution of (3.74) ϵ_2 and values $N_0 > 0$, $\delta_0 > 0$ and $0 < \epsilon < 1$ are chosen.

General algorithm step: suppose that the point x_k and numbers N_k and δ_k have already been constructed.

1. We construct the sets $I_\delta^-(x_k)$ and $I_\delta^0(x_k)$ and formulate problem (3.74).

2. We solve problem (3.74). The Lagrange multipliers corresponding to the solution are denoted by u_k^i, $i \in I_{\delta_k}^-(x_k) \cup I_{\delta_k}^0(x_k)$. If problem (3.74) is inconsistent, we set

   $$x_{k+1} = x_k,\ \ \delta_{k+1} = \delta_k/2,\ \ N_{k+1} = N_k$$

   and return to stage 1.

3. If the solution of problem (3.74) has been obtained and

$$N_k \geq \sum_{i\in I^-_{\delta_k}(x_k)} u^i_k + \sum_{i\in I^0_{\delta_k}(x_k)} |u^i_k|$$

then $N_{k+1} = N_k$.

Otherwise

$$N_{k+1} = 2\left[\sum_{i\in I^-_{\delta_k}(x_k)} u^i_k + \sum_{i\in I^0_{\delta_k}(x_k)} |u^i_k|\right].$$

4. We then compute the vector for the direction of movement

$$p_k = -f'_0(x_k) - \sum_{i\in I^-_{\delta_k}(x_k)} u^i_k f'_i(x_k) - \sum_{i\in I^0_{\delta_k}(x_k)} u^i_k f'_i(x_k). \tag{3.75}$$

5. If

$$\|p_k\|^2 \leq \epsilon_1 \tag{3.76}$$

we move to stage 7.

6. If (3.76) is not satisfied, we set

$$x_{k+1} = x_k + \alpha_k p_k, \quad \delta_{k+1} = \delta_k$$

where the step α_k is chosen to equal $1/2^{q_0}$ where q_0 is the first integer $q = 0, 1, \ldots$ for which the inequality

$$\Phi_{N_{k+1}}\left(x_k + \frac{1}{2^q}p_k\right) \leq \Phi_{N_{k+1}}(x_k) - \frac{1}{2^q}\epsilon\|p_k\|^2 \tag{3.77}$$

is satisfied. The algorithm now returns to stage 1.

7. Conclusion.

From practical experience of the operation of this algorithm, we deduce that δ_0 should be chosen to be sufficiently large that all the constraints are covered. However, here, we must take into account another factor relating to solvability which arises in the case of this problem (3.74). If the problem is solvable all the constraints are considered at once, the active ones are quickly identified and $u(x_k) \to u_*$, $x_k \to x_*$. Otherwise, we have to reduce δ and begin work again for each reduction. If δ_0 is large, this may be expensive in terms of machine time.

A choice of δ_0 in the form

$$\delta_0 = F(x_0) + \epsilon^1$$

for some number $\epsilon^1 > 0$ is highly recommended. For such a choice of δ_0 all the constraints which are strongly violated are immediately taken into account and the point x_k quickly begins to satisfy all the constraints of the problem. For example, ϵ^1 may be chosen to be the ϵ occurring in (3.77). However, it is better to choose it with reference to a specific problem or a classical problem.

Suppose that the constraints (3.72) include simple constraints on the coordinates of the vector x of the type

$$a^j \leq x^j \leq b^j, \ \ j = 1, \ldots, n \tag{3.78}$$

where a^j and b^j are arbitrary numbers. In practice, in real problems, such constraints are introduced based on physical considerations and their violation often leads to various types of failure. At the start of the iteration process of the linearization method, when the set $I_\delta(x)$ (2.3) still bears little resemblance to the set of active constraints $I_0(x_*) = I_*$, it is possible to output a point x_k violating the constraints. Thus, in the presence of constraints of type (3.78), it is recommended that $I_\delta(x)$ be taken to include a priori all constraints of this form and that the technique for constructing the set $I_\delta(x)$ be applied to the remaining constraints.

Despite the fact that the symmetry of the matrix A in problem (3.74) means that we only have to store half of this matrix in the computer, most of the computer memory used is used to store this. It is clear from the form of the elements of the matrix A that, in the presence of constraints of type (3.78), it takes on a block structure. All the blocks except one consist of elements, which can be determined without requiring a computer to calculate the scalar products $(f_i'(x), f_j'(x))$. This means that only a small part of this matrix (or none of it at all) need be stored in the computer memory. If the vector x is of a large dimension and constraints (3.78) are imposed on all its components, this leads to a significant memory saving.

Here is another remark about the relationship between the values of ϵ_1 and ϵ_2 used in the algorithm. The accuracy ϵ_2 with which (3.74) is solved should be set to be one or more orders greater than the accuracy ϵ_1 with which the main problem is solved, since insufficiently accurate solution of problem (3.74) may not lead to results.

3.3.2 The Accelerated Linearization Method

Unlike the accelerated version of algorithm studied in Section 2.3, the algorithm given here may be more difficult to visualize but more amenable to practical use. We assume that the reader is familiar with Section 3.3.1, since in studying the algorithm we may omit certain details, for example, relating to the choice of the parameters N_k and δ_k. We introduce the set

$$I(x) = \{i \in I_\delta^-(x) \cup I_\delta^0(x) | f_i(x) = 0\}$$

In the description of the accelerated version of the algorithm in Section 2.3, we showed that the iterative sequence $\{x_k\}$ was divided into segments in which updating from x_k to x_{k+1} took place either according to the formula

$$x_{k+1} = x_k + y_k \tag{3.79}$$

where y_k is a solution of the system of equations

$$\begin{aligned} A(x_k, h_k)y + (f_0'(x_k))^* + \sum_{i \in I(x_k)} v^i (f_i'(x_k))^* &= 0 \\ f_i'(x_k)y + f_i(x_k) &= 0, \quad i \in I(x_k), \end{aligned} \tag{3.80}$$

or according to the formula of the linearization method with the step chosen using formula (3.77). Here, the step from x_k to x_{k+1} uses formula (3.79) only when the condition

$$\|p(x_{k+1})\| \le \gamma \|p_k\| \tag{3.81}$$

is satisfied, where $0 < \gamma < 1$ is as in Section 2.3.

We define a quantity $\mathcal{F}$ which takes the values:

$$\mathcal{F} = \begin{cases} 1 & \text{if the system (3.80) is solvable;} \\ 2 & \text{if the system (3.80) is solvable and} \\ & \text{the condition (3.81) is satisfied;} \\ 0 & \text{if the system (3.80) is unsolvable or} \\ & \text{the condition (3.81) is not satisfied.} \end{cases}$$

Let us now define the algorithm for the accelerated method. Suppose that $N > 0$, $\delta > 0$, an initial approximation x_0 and numbers $0 < \gamma < 1$, $0 < \epsilon < 1$, $0 < q_0 < 1$, $h > 0$, ϵ_1 and ϵ_2 are chosen. We set $C_0 = +\infty$, $\alpha_0 = +\infty$ and $\mathcal{F} = 0$.

(The value $+\infty$ is taken to be the largest possible number permitted by the computer on which the algorithm is implemented.)

General algorithm step: suppose that x_k, C_k and q_k have already been constructed.

1. We solve problem (3.74) for $x = x_k$. We obtain $u_k^i = u^i(x_k)$, $i \in I_{\delta_k}^-(x_k) \cup I_{\delta_k}^0(x_k)$. We compute the vector $p_k = p(x_k)$.

2. If $\mathcal{F} = 1$ we move to 8.

3. We check whether or not the point x_k is a solution. If $\|p_k\|^2 \le \epsilon_1$ then we move to 10.

4. If $\mathcal{F} = 2$ we move to 7.

5. If $\alpha_{k-1}\|p_k\| > q_k$, then $q_{k+1} = q_k$ and we move to 9.

6. If $\|p_k\| > C_k$ then we set $q_{k+1} = q_k$ and move to 9.

7. We set $h_k = \min\{h, \|p_k\|\}$ and solve the system of equations (3.80) relating to $\mathcal{F}$. If the system does not have a solution then we set $q_{k+1} = q_k$, $C_{k+1} = \gamma\|p_k\|$ and $\mathcal{F} = 0$ and move to 9. Otherwise, we set

$$\overline{x} = x_k + y_k, \quad \mathcal{F} = 1 \tag{3.82}$$

and return to 1 to compute $p(\overline{x})$.

8. If

$$\|p(\overline{x})\| \le \gamma \|p_k\| \tag{3.83}$$

then we set

$$x_{k+1} = \overline{x}, \quad C_{k+1} = \gamma \|p_k\|, \quad q_{k+1} = q_k, \quad \mathcal{F} = 2$$

and move to 3. Otherwise, we set

$$C_{k+1} = \gamma \|p_k\|, \quad q_{k+1} = q_k/2, \quad \mathcal{F} = 0$$

and move to the next step.

9. We set $\alpha = 1$ and divide it by two until the following inequality is satisfied.

$$f_0(x_k + \alpha p_k) + NF(x_k + \alpha p_k) \le f_0(x_k) + NF(x_k) - \alpha \epsilon \|p_k\|^2. \tag{3.84}$$

We set

$$x_{k+1} = x_k + \alpha p_k \tag{3.85}$$

and return to 1.

10. Conclusion.

From the algorithm thus constructed, it is clear that, at the start of the iteration process when the point x_k is still far away from x_* only the linearization method is used. As soon as the conditions $\alpha_{k-1}\|p_k\| \le q_k$, $0 < q_k < 1$, $q_k \to 0$ are satisfied, a check is included to test whether the conditions for operation according to formula (3.82) are satisfied. The theoretical calculations carried out in Section 2.3, which provide for stepping according to (3.82) are only valid in a local neighbourhood of the point x_*. Thus, the introduction into the algorithm of the parameter q_k for practical purposes is fully justified, since otherwise much unnecessary work will be carried out. When choosing q_0, it is possible to take into account, for example, the given accuracy ϵ_1 for solution of the original problem. Thus, in calculations in which this accuracy is given to be in the range 10^{-4}–10^{-7}, the parameter q_0 was set equal to 0.1 and good results were obtained.

On the other hand, it is clear from the algorithm that after the first successful step using formula (3.82) the next step will again use this formula: of course a check is made to ensure that (3.83) is satisfied.

In our calculations, transition from formula (3.82) to formula (3.85), and conversely, took place mainly when formula (3.82) was first introduced into the computation. The last iteration of the process used formula (3.82). The value of γ was taken to be close to one. Good convergence was obtained, even for $\gamma = 1$.

The system of equations (3.80) may be solved using any method designed to solve systems of linear algebraic equations, for example, Gauss's method.

3.3.3 Examples of Calculations

We give a number of examples of problems solved using the algorithms described in Sections 3.3.1 and 3.3.2.

Example 1 Minimize

$$f_0(x) = (x_1 - 1)^2 + (x_1 - x_2)^2 + (x_2 - x_3)^3 + (x_3 - x_4)^4 + (x_4 - x_5)^4$$

subject to the constraints

$$\begin{aligned} f_1(x) &= x_1 + x_2^2 + x_3^3 = 2 + 3\sqrt{2} \\ f_2(x) &= x_2 - x_3^2 + x_4 = -2 + 2\sqrt{2} \\ f_3(x) &= x_1 x_5 = 2 \end{aligned}$$

Thus, we have to minimize a nonlinear function of five variables with three nonlinear equality-type constraints. This problem was solved from different initial values using the linearization method and the accelerated method.

The following points were used as initial starting points:

$$\begin{aligned} x_{0_1} &= (-1, 3, -0.5, -2, -3) \\ x_{0_2} &= (-1, 2, 1, -2, -2) \\ x_{0_3} &= (-2, -2, -2, -2, -2) \\ x_{0_4} &= (1, 1, 1, 1, 1) \end{aligned}$$

Corresponding to these starting points we obtained the following solutions:

$$\begin{aligned} x_{*_1} &= (-0.7034, 2.636, -0.09636, -1.798, -2.843) \\ x_{*_2} &= (-1.273, 2.410, 1.195, -0.1542, -1.571) \\ x_{*_3} &= (-2.791, -3.004, 0.2054, 3.875, -0.7166) \\ x_{*_4} &= (1.117, 1.220, 1.538, 1.973, 1.791). \end{aligned}$$

Table 1 shows the results of calculations by both methods and may be used to compare the characteristics of these methods.

Values $q_0 = 0.1$ and $\gamma = 1$ were assigned. One computation of the functions is taken to mean one computation of the function to be minimized $f_0(x)$ and the functions $f_i(x)$, $i \in I_\delta(x)$.

From Table 1 it is clear that the accelerated method is particularly effective when the linearization method moves with small steps, since reducing the step is costly in terms of resources. Each reduction requires the computation of the functions to check condition (3.84). Starting from the point x_{0_4}, the linearization method seems quite effective.

In the accelerated method, the main expense in computing the functions relates to the construction of the matrix $A(x, h)$. To construct the $n \times n$ matrix $A(x, h)$ once requires $n(1 + 2n)$ computations of the functions. In this example, for the initial point x_{0_1} using the accelerated method, the functions had to be evaluated 220 times in order to construct the 5×5 matrix $A(x, h)$ four times, whereas for a total of 19 iterations of the method it was only necessary to compute the functions 19 times.

From this, it becomes clear that it is advisable to introduce the parameter q_k into the algorithm which initially provides for operation of the process

Table 1

Method	x_0	x_*	$f(x_*)$	$\|p\|$
Linearization	x_{0_1}	x_{*1}	44.02	$1.24 \cdot 10^{-4}$
	x_{0_2}	x_{*2}	27.87	$2.33 \cdot 10^{-5}$
	x_{0_3}	x_{*3}	607.04	$1.74 \cdot 10^{-3}$
	x_{0_4}	x_{*4}	0.0293	$7.74 \cdot 10^{-6}$
Accelerated	x_{0_1}	x_{*1}	44.02	$5.12 \cdot 10^{-5}$
	x_{0_2}	x_{*2}	27.87	$1.27 \cdot 10^{-5}$
	x_{0_3}	x_{*3}	607.04	$1.88 \cdot 10^{-6}$
	x_{0_4}	x_{*4}	0.0293	$4.04 \cdot 10^{-5}$

Total iterations	Iterations via (3.82)	Surplus iterations via (3.82)	Computations of functions	Computations of gradients	Step to next iteration
481	–	–	3516	482	$1.3 \cdot 10^{-13}$
399	–	–	2623	400	$9.09 \cdot 10^{-13}$
605	–	–	6506	606	$9.09 \cdot 10^{-13}$
21	–	–	49	22	1
19	4	0	239	20	1
29	3	0	258	30	1
112	8	5	1313	118	1
11	3	0	176	12	1

according to the pure linearization method, decreases the probability of surplus computations, and implements stepping by formula (3.82) when it may still be invalid.

From Table 1 it is also clear that the cost of computing the gradients of the functions in the accelerated method is in all cases considerably less than in the linearization method.,

Let us consider another example of minimization of a nonlinear function with two nonlinear constraints [43]. One constraint is an equality, the other is an inequality.

Example 2 Minimize

$$f_0(x) = 0.5(x_1 + x_2)^2 + 50(x_2 - x_1)^2 + x_3^2$$

subject to the constraints

$$\begin{aligned} f_1(x) &= (x_1 - 1)^2 + (x_2 - 1)^2 + (x_3 - 1)^2 - 1.5 \leq 0 \\ f_2(x) &= \sin(x_1 + x_2) - x_3 = 0. \end{aligned}$$

The initial point is $x_0 = (-1, 4, 5)$, The minimum occurs at the point $x_* = (0.229, 0.229.0.442)$ and $f_0(x_*) = 0.3$, This problem was solved by the linearization method and by the accelerated method. To achieve an accuracy of

$\|p\|$ equal to 10^{-5}, 740 iterations of the linearization method were required and 27 iterations of the accelerated method.

For $\gamma = 1$ and $q_0 = 0.1$, formula (3.82) was used three times. The first time transition from (3.85) to (3.82) was premature. Two successive steps used (3.82).

Since the difference in the numbers of iterations for these two methods is very large, the accelerated method clearly appears more effective here.

One feature of the next example [18] is that the matrix of second derivatives of the Lagrange function with respect to x has a negative eigenvalue -136 and all positive eigenvalues less than 2; in other words, it is not positive definite.

Example 3 The goal function is

$$f_0(x) = e^{x_1 x_2 x_3 x_4 x_5} - 0.5(x_1^3 + x_2^3 + 1)^2.$$

The equality constraints are

$$\begin{aligned} f_1(x) &= x_1^2 + x_2^2 + x_3^2 + x_4^2 + x_5^2 - 10 = 0 \\ f_2(x) &= x_2 x_3 - 5x_4 x_5 = 0 \\ f_3(x) &= x_1^3 + x_2^3 + 1 = 0. \end{aligned}$$

The initial point is $x_0 = (-2, 2, 2, -1, -1)$. The minimum is $x_* = (-1.717, 1.596, 1.827, -0.7636, -0.7636)$. The value of the goal function at x_* is 0.05395.

To achieve an accuracy of $\|p\|$ of 10^{-4}, the linearization method required 32 iterations and the accelerated method 6 iterations, of which the last 4 used formula (3.82).

We now give further examples of problems solved by the linearization method and its variants which are described and illustrated in [45, p. 416–424]. A number of problems are considered involving the same goal function but with different sets of inequality constraints. The goal function depends on two variables and has a peak and a saddle point in the given range of values of these variables. The peak of this goal function is at the point with coordinates $x_1 = 81.154841$, $x_2 = 69.135588$ and $f(x_*) = 61.9059345$. The conditional maximum is at the point with coordinates $x_1 = 75.000000$, $x_2 = 65.000000$ and the value of the function at this point is 58.903436.

The goal function has the form

$$\begin{aligned} f(x) = {} & B_1 + B_2 x_1 + B_3 x_1^2 + B_4 x_1^3 + B_5 x_1^4 + B_6 x_2 \\ & + B_7 x_1 x_2 + B_8 x_1^2 x_2 + B_9 x_1^3 x_2 + B_{10} x_1^4 x_2 \\ & + B_{11} x_2^2 + B_{12} x_2^3 + B_{13} x_2^4 + B_{14}(1/(x_2 + 1)) \\ & + B_{15} x_1^2 x_2^2 + B_{16} x_1^3 x_2^2 + B_{17} x_1^3 x_2^3 + B_{18} x_1 x_2^2 \\ & + B_{19} x_1 x_2^3 + B_{20} e^{0.0005 x_1 x_2} \end{aligned}$$

where

$$
\begin{array}{lll lll}
B_1 & = & 75.1963666677 & B_{11} & = & 0.2564581253 \\
B_2 & = & -3.8112755343 & B_{12} & = & -0.0034604030 \\
B_3 & = & 0.1269366345 & B_{13} & = & 0.0000135139 \\
B_4 & = & -0.0020567665 & B_{14} & = & -28.1064434908 \\
B_5 & = & 0.0000103450 & B_{15} & = & -0.0000052375 \\
B_6 & = & -6.8306567613 & B_{16} & = & -0.0000000063 \\
B_7 & = & 0.0302344793 & B_{17} & = & 0.0000000007 \\
B_8 & = & -0.0012813448 & B_{18} & = & 0.0003405462 \\
B_9 & = & 0.0000352599 & B_{19} & = & -0.0000016638 \\
B_{10} & = & -0.0000002266 & B_{20} & = & -2.8673112392
\end{array}
$$

The inequality constraints which were used to construct the individual problems were:

$$
\begin{aligned}
f_1(x) &= -x_1 \leq 0 \\
f_2(x) &= -x_2 \leq 0 \\
f_3(x) &= x_1 - 95 \leq 0 \\
f_4(x) &= x_2 - 75 \leq 0 \\
f_5(x) &= 700 - x_1 x_2 \leq 0 \\
f_6(x) &= x_1 - 75 \leq 0 \\
f_7(x) &= x_2 - 65 \leq 0 \\
f_8(x) &= 5(x_1/25)^2 - x_2 \leq 0 \\
f_9(x) &= 5(x_1 - 55) - (x_2 - 50)^2 \leq 0 \\
f_{10}(x) &= 54 - x_1 \leq 0 \\
f_{11}(x) &= (x_1 - 45) - (3/2)(x_2 - 45) \leq 0 \\
f_{12}(x) &= (4/25)(x_2 - 40) - x_1 + 35 \leq 0.
\end{aligned}
$$

Table 2 shows the indices of the constraints used in the five individual problems. All the problems involved maximization of $f_0(x)$ subject to the given constraints.

Table 2

Problem number	Indices of constraints
I	1,2,3,4,5
II	5,6,7,8
III	5,6,7,8,9
IV	5,6,7,8,9,10
V	5,6,7,8,9,11,12

The pure linearization method was quite effective in solving all these problems and Table 3 gives the results of the calculations using this method.

In solving these problems the linearization method used a step of one in almost all iterations. Thus, the difference in the number of iterations for the accelerated method here was not as marked as in the previous examples when

Table 3

Problem number	Initial point x_0	x_*	$f_0(x_*)$	$\|p\|$	Number of iterations
I	(95,10)	(81.1548,69.1356)	61.9059	$5.47 \cdot 10^{-6}$	74
II	(31,48)	(75,65)	58.9034	$2.50 \cdot 10^{-9}$	22
III	(31,48)	(75,65)	58.9034	$4.03 \cdot 10^{-9}$	47
IV	(68.8,31.2)	(75,65)	58.9034	$5.06 \cdot 10^{-19}$	51
V	(68.8,31.2)	(75,65)	58.9034	$6.06 \cdot 10^{-17}$	35

the step in the linearization method was very small. For example, beginning at the unfeasible point $x_0 = (95, 10)$ in the first problem a solution was obtained using the accelerated method after 56 iterations, where formula (3.82) was only used in the last iteration, while the linearization method took 55 iterations.

Many of the problems described in [45] were also solved using the methods described. The behaviour of the methods corresponds to the pattern described in the previous examples.

Here, we shall not go into further details of the list of these examples, but give an example of the calculations involved in an actual problem. This problem arises in a national economic context and relates to the planning of systematic use of water resources in the basin of the R.Dnieper [22]. The problem has 136 constraints, of which 112 are simple constraints on the coordinates and 24 are nonlinear equality constraints.

Let us formulate the problem. Maximize

$$\begin{aligned} f_0(x) &= \sum_{i=1}^{12}(-2.188 + 19.95x_{24+i} + 0.07656x_i) \\ &\quad + \sum_{i=1}^{12}(11.56 - 24.89x_{36+i} - 0.7135x_{12+i} + 2.155x_{36+i}x_{12+i}) \end{aligned}$$

subject to the constraints

$$\begin{aligned} &f_1(x)\dots f_{24}(x): \quad 51.2 \le x_i \le 51.4 \\ &f_{25}(x)\dots f_{48}(x): \quad 15 \le x_{12+i} \le 16.1 \\ &f_{49}(x)\dots f_{72}(x): \quad 0.4 \le x_{24+i} \le 4.6 \\ &f_{73}(x)\dots f_{96}(x): \quad 0.5 \le x_{36+i} \le 4.8 \\ &f_{97}(x)\dots f_{104}(x): \quad 0 \le x_{48+j} \le 0.7 \\ &f_{105}(x)\dots f_{112}(x): \quad 0 \le x_{52+j} \le 0.7 \\ &f_{112+j}(x) = W(x_j, x_{24+j}) + c_j - 2.68x_{24+j} - W(x_{j-1}, x_{24+j-1}) = 0 \\ &f_{116+j}(x) = W(x_{4+j}, x_{28+j}) + c_{4+j} \\ &\qquad - 2.68x_{28+j} - W(x_{3+j}, x_{27+j}) - 2.68x_{48+j} = 0 \\ &f_{120+j}(x) = W(x_{8+j}, x_{32+j}) + c_{8+j} \\ &\qquad - 2.68x_{32+j} - W(x_{7+j}, x_{31+j}) = 0 \end{aligned}$$

$$
\begin{aligned}
f_{124+k}(x) &= W(x_{12+k}, x_{36+k}) + c_{12+k} \\
&\quad - 2.68(x_{24+k} + x_{36+k}) - W(x_{11+k}, x_{35+k}) = 0 \\
f_{128+k}(x) &= W(x_{16+k}, x_{40+k}) + c_{16+k} \\
&\quad - 2.68(x_{28+k} + x_{40+k}) - W(x_{15+k}, x_{39+k}) - 2.68 x_{52+k} = 0 \\
f_{132+k}(x) &= W(x_{20+k}, x_{44+k}) + c_{20+k} \\
&\quad - 2.68(x_{32+k} + x_{44+k}) - W(x_{19+k}, x_{43+k}) = 0 \\
i &= 1, \dots, 12, \quad j = k = 1, \dots, 4
\end{aligned}
$$

Table 4

		a_1	a_2	a_3	a_4	a_5
j	$1 \dots 4$	34.547	−0.55878	8.05339	−0.02252	−0.29316
k	$1 \dots 4$	20.923	−4.22088	1.42061	−0.41040	−0.15082

Table 4 (continuation)

		a_6	a_7	a_8	a_9	a_{10}
j	$1 \dots 4$	−0.013521	0.00042	0.00267	0.000281	0.0000032
k	$1 \dots 4$	0	−0.00826	0.00404	0.000168	−0.000038

Table 5

i	1	2	3	4	5	6
c_i	5.61	4.68	1.62	1.8	2.13	2.1
c_{12+i}	−0.02	−0.01	−0.16	−0.47	−0.75	−0.94

Table 5 (continuation)

i	7	8	9	10	11	12
c_i	1.99	2.02	2.14	2.15	2.36	2.63
c_{12+i}	−0.93	−0.99	−0.42	−0.07	0.04	−0.06

Table 6

i	1	2	3	4	5
x_{*_i}	51.40	51.40	51.40	51.40	51.40
$x_{*_{12+i}}$	16.10	16.10	16.10	16.10	16.10
$x_{*_{24+i}}$	2.009	1.751	0.6233	0.6709	0.7928
$x_{*_{36+i}}$	1.502	1.740	0.5835	0.5000	0.5096
$x_{*_{48+i}}$	$-1.49 \cdot 10^{-7}$	$-2.7 \cdot 10^{-7}$	$1.95 \cdot 10^{-7}$	$1.71 \cdot 10^{-8}$	$1.12 \cdot 10^{-7}$

Table 6 (continuation)

i	6	7	8	9	10	11	12
x_{*_i}	51.38	51.38	51.37	51.20	51.20	51.20	51.20
$x_{*_{12+i}}$	16.03	15.92	15.80	15.89	15.64	15.00	15.00
$x_{*_{24+i}}$	0.7871	0.7431	0.7551	0.8245	0.8026	0.8792	0.9796
$x_{*_{36+i}}$	0.5000	0.5000	0.5000	0.5852	1.002	1.482	0.9712
$x_{*_{48+i}}$	$1.71 \cdot 10^{-7}$	$-7.84 \cdot 10^{-8}$	$-1.74 \cdot 10$				

where

$$\begin{aligned} W(x,y) &= a_1 + a_2x + a_3y + a_4x^2 + a_5xy + a_6y^2 \\ &\quad + a_7x^3 + a_8x^2y + a_9xy^2 + a_{10}y^3 \end{aligned}$$

For $j = k = 1$, $W(x_0, x_{24}) = W(50.82, 2)$ and $W(x_{12}, x_{36}) = W(15.5, 2.3)$. The values of the coefficients a_i, $1 \le i \le 10$ for $W(x,y)$ are given in Table 4 and the values of the c_i, $1 \le i \le 24$ are given in Table 5.

The problem was solved by the linearization method. Beginning at the point x_0, the point x_* was found after 13 iterations with $\|p\| = 8.139 \cdot 10^{-4}$. The point x_0 was the following: $x_{1_0} \ldots x_{12_0} = 51.35$, $x_{13_0} \ldots x_{24_0} = 15.5$, $x_{25_0} \ldots x_{36_0} = 2.5$, $x_{37_0} \ldots x_{48_0} = 2.6$, $x_{49_0} \ldots x_{56_0} = 0.3$. The coordinates of the point x_* are given in Table 6. The value of the goal function at the point x_* was $f(x_*) = 36.4138$. The iterative process of the linearization method used a step $\alpha_k = 1$, $1 \le k \le 13$. All the calculations given here were carried out on a 'BESM 6' computer.

Appendix: Comments on the Literature

Since the literature on necessary conditions for extrema, duality theory in convex programming, minimax problems and numerical techniques for solving various extremal problems includes thousands of references, it is impossible to undertake any form of comparative analysis here. Thus, we shall only refer to books and papers mentioned in the text which are directly related to the subject matter of this book.

As far as the theory of necessary conditions for extrema and that of duality in convex programming are concerned, we based our study in Section 1.2 on the books by A.D. Ioffe and V.M. Tikhomirov [7], B.N. Pshenichnyj [23,28], R.T. Rockafellar [33], and I. Ekeland and R. Temam [47]. A detailed theoretical analysis of penalty function techniques is given in the book by K. Grosman and A.A. Kaplan [2] which includes further references to the literature. However, we based our study of the conditions under which the minima of penalty functions coincide with the solutions of the original problem (Section 1.2.7) on the paper by F. Clarke [9].

Quadratic programming problems are fundamental to the use of the linearization method. The speed and economy (from the point of view of the computations and the computer storage involved) with which they are solved determines the effectiveness of the method, particularly as far as large problems are concerned. In this connection, Section 1.3 was based on a technique for solving quadratic programming problems which generalizes the simplex method of linear programming (see G.B. Dantzig [3]). Here, we considered the need for economic use of computer resources when working with sparse matrices, and used the approaches developed by B.A. Murtagh and M.A. Saunders [11] together with a multiplicative representation of inverse matrices. This representation is not the only one possible nor is it always optimal. More details of this may be found in R. Tewarson [36]. In addition to the methods for solving quadratic programming problems studied in Section 1.3, there are also other methods which are described in the paper by V.A. Daugavet [5], in the books by H.P. Künzi and W. Krelle [10] and B.N. Pshenichnyj and Yu.M. Danilin [28] and in the collected papers [46]. It should be said that most of these methods carry over to the problem of minimizing an arbitrary function subject to linear constraints, as is shown in the aforementioned articles [11,46] and in the paper by B.N. Pshenichnyj and I.F. Ganzhela [27]. A very detailed study of conjugate gradient methods in relation to the minimization of quadratic functions is given in [28] and in the book by E. Polak [19].

The linearization method in the form described in this book was first studied by B.N. Pshenichnyj in [24] and, in relation to equations and inequalities, in [25]. Modifications of it, with differences in the auxiliary problem, are considered by U.M. Garcia-Palomares and O.L. Mangasarian [1], S.M. Robinson [31, 32], and S.P. Han [39]. Quite rightly, these only consider local convergence. Later, in the papers [40–42], S.P. Han considered questions of global convergence. The papers by V.M. Panin [13–16] and M. Powell [17,18] are devoted to the same problem area. We note that in these articles, a superlinear rate of convergence was achieved as a result of the (explicit and implicit) use of the second derivatives of the Lagrange function in the auxiliary problem. However, in fact, this required the matrix of second derivatives of the Lagrange functions to be strictly positive definite. This shortcoming was overcome by B.N. Pshenichnyj and L.A. Sobolenko [30].

Resolution of minimax problems by the linearization method (other techniques for solving these problems are considered in the book by V.F. Dem'yanov and V.N. Malozemov [6]) was studied in the book by B.N. Pshenichnyj and Yu.M. Danilin [28] and in the paper by B.N. Pshenichnyj [26]. The fact that the linearization method guarantees quadratic convergence for Chebyshev points was proved by V.A. Daugavet and V.N. Malozemov [4]. Solutions of various generalized problems using minimax methods which are modifications of the linearization method were considered by K. Kiwiel [8], V.M. Panin [13,14], S.P. Han [42] and R. Fletcher [38].

All the calculations using the linearization method given in Section 3.3 were carried out by L.A. Sobolenko. In addition to problems from the literature and a number of real problems were also solved. It is impossible to describe all this material in the book and thus Section 3.3 only includes examples which are, in the author's opinion, in some sense characteristic. In the absence of a standard against which to assess the results of calculations given in the literature, it is difficult to make an accurate comparison; however, it is safe to say that the results of numerical computations using the linearization method or its accelerated modification are no worse than those of any other method.

References

1. Garcia-Palomares, U.M. and Mangasarian, O. (1976): Superlinearly convergent quasi-Newton methods for nonlinearly constrained optimization problems. Math. Program. **11** (1976) 1–13
2. Grosman, K. and Kaplan, A.A. (1981): Nonlinear Programming Based on Unconstrained Optimization. Nauka, Novosibirsk
3. Dantzig, G.B. (1962): Linear Programming and its Extensions. Princeton University Press, Princeton NJ
4. Daugavet, V.A. and Malozemov, V.N. (1981): Quadratic rate of convergence of a linearization method for solving discrete minimax problems. Zh. Vychisl. Mat. Mat. Fiz. **21** 4 (1981) 835–843
5. Daugavet, V.A. (1981): Modification of a method of Wulf. Zh. Vychisl. Mat. Mat. Fiz. **21** 2 (1981) 504–508
6. Dem'yanov, V.F. and Malozemov, V.N. (1972): Introduction to Minimax. Nauka, Moscow. English transl.: John Wiley and Sons (1974)
7. Ioffe, A.D. and Tikhomirov, V.M. (1974): Theory of Extremal Problems. Nauka, Moscow. English transl.: Academic Press (1979)
8. Kiwiel, K.C. (1981): A globally convergent quadratic approximation for inequality constrained minimax problems. IIASA Collab. Proc. Ser. **81-9** (1981) 1–24
9. Clarke, F. (1976): A new approach to Lagrange multipliers. Math. Oper. Res. **1** 2 (1976) 165–174
10. Künzi, H.P. and Krelle, W. (1962): Nichtlineare Programmierung. Springer-Verlag
11. Murtagh, B.A. and Saunders, M.A. (1978): Large-scale linearly constrained optimization. Math. Program. **14** (1978) 41–72
12. Ostrovskij, A.M. (1963): Solution of Equations and Systems of Equations. IL, Moscow. English transl.: Academic Press (1966)
13. Panin, V.M. (1980): The linearization method for the discrete minimax problem. Kibernetika **3** (1980) 86–90
14. Panin, V.M. (1981): The linearization method for the continuous minimax problem. Kibernetika **2** (1981) 75–78
15. Panin, V.M. (1981): Methods for solving convex programming problems. Zh. Vychisl. Mat. Mat. Fiz. **21** 2 (1981) 315–328
16. Panin, V.M. (1982): Global convergence of the damped Newtonian method in convex programming problems. Dokl. Akad. Nauk. SSSR **261** 4 (1982) 811–814
17. Powell, M. (1977): The convergence of variable matrix methods for nonlinearly constrained optimization calculations. Dept. Appl. Math. and Theoret. Phys.
18. Powell, M. (1978): Algorithms for nonlinear constraints that use Lagrangian functions. Math. Program. **14** (1978) 224–228

19. Polak, E. (1971): Computational Methods in Optimization. A Unified Approach. Academic Press, New York, London
20. Polyak, B.T. and Tret'yakov, N.V. (1972): An iterative linear programming method and its economic interpretation. Ehkon. Mat. Metody **VIII** 5 (1972) 740–751
21. Polyak, B.T. and Tret'yakov, N.V. (1973): A penalty function method for conditional extremum problems. Zh. Vychisl. Mat. Mat. Fiz. **13** 1 (1973) 34–46
22. Popkov, N.V., Tulupchuk, Yu.M. and Khilyuk, L.F. (1978). Three-criterion control of the water management system of the Dnieper basin. Avtomatika **2** (1978) 44–53
23. Pshenichnyj, B.N. (1969): Necessary Conditions for an Extremum. Nauka, Moscow. English transl.: Marcel Dekker Inc., New York (1971)
24. Pshenichnyj, B.N. (1970): Algorithm for a general mathematical programming problem. Kibernetika **5** (1970) 120–125
25. Pshenichnyj, B.N. (1970): Newton's method for solving systems of equations and inequalities. Mat. Zametki **8** 5 (1970) 635–640
26. Pshenichnyj, B.N. (1978). Nonsmooth optimization and nonlinear programming. In Lemarechal, V.C. and Mifflin, R. (eds.) (1978): Nonsmooth Optimization. Pergamon Press
27. Pshenichnyj, B.N. and Ganzhela, I.F. (1970): An algorithm for solving convex programming problems with linear constraints. Kibernetika **3** (1970) 81–85
28. Pshenichnyj, B.N. and Danilin, Yu.M. (1978): Numerical Methods in Extremal Problems. Mir, Moscow
29. Pshenichnyj, B.N. (1980): Convex Analysis and Extremal Problems. Nauka, Moscow
30. Pshenichnyj, B.N. and Sobolenko, L.A. (1980): Acceleration of the convergence of the linearization method for problems of conditional minimization. Zh. Vychisl. Mat. Mat. Fiz. **20** 3 (1980) 605–614
31. Robinson, S.M. (1972): A quadratically convergent algorithm for general nonlinear programming problems. Math. Program. **3** (1972) 145–156
32. Robinson, S.M. (1974): Perturbed Kuhn–Tucker points and rates of convergence for a class of nonlinear programming algorithm. Math. Program. **4** (1974) 1–16
33. Rockafellar, R.T. (1970): Convex Analysis. Princeton University Press, Princeton NJ
34. Rockafellar, R.T. (1976): Augmented Lagrangians and applications of the proximal point algorithm in convex programming. Math. Oper. Res. **1** 2 (1976) 97–115
35. Tret'yakov, N.V. (1973): Penalty function method for convex programming problems, Ehkon. Mat. Metody **IX** 3 (1973) 525–540
36. Tewarson, R. (1973): Sparse Matrices. Academic Press
37. Fedorenko, R.P. (1978): Approximate Solution of Optimal Control Problems. Nauka, Moscow
38. Fletcher, R. (1981): Second order corrections for nondifferentiable optimization. Numer. Anal. Rep. **50** (1981) 1–30
39. Han, S.P. (1976): Superlinearly convergent variable metric algorithms for general nonlinear programming problems. Math. Program. **11** (1976) 263–282

40. Han, S.P. (1977): A globally convergent method for nonlinear programming. J. Optimization Theory Appl. **22** (1977) 297–309.
41. Han, S.P. (1977): Dual variable metric method for constrained optimization, SIAM J. Control Optimization **15** (1977) 546–565
42. Han, S.P. (1981): Variable metric method for minimizing a class of nondifferentiable functions. Math. Program. **20** (1981) 1–13
43. Khenkin, Eh.I. and Volynskij, M.Z. (1976): Search algorithm for solving a general mathematical programming problem, Zh. Vychisl. Mat. Mat. Fiz. **10** 1 (1976) 61–71
44. Hestenes, M.R. (1969): Multiplier and gradient methods, J. Optimization Theory Appl. **4** (1969) 303–320
45. Himmelblau, D. (1972): Applied Nonlinear Programming. McGraw-Hill
46. Gill, P., Murray, W. and Wright, M. (1981): Practical Optimization. Academic Press, New York, London
47. Ekeland, I. and Temam, R. (1976): Convex Analysis and Variational Problems. North-Holland Elsevier, Amsterdam

SPRINGER SERIES IN COMPUTATIONAL MATHEMATICS

Volume 22: **B. N. Pshenichnyj**

The Linearization Method for Constraint Optimization

1994. ISBN 3-540-57037-3

Volume 21: **R. Hammer, M. Hocks, U. Kulisch, D. Ratz**

Numerical Toolbox for Verified Computing I

Basic Numerical Problems

1993. ISBN 3-540-57118-3

Volume 20: **F. Stenger**

Numerical Methods Based on Sinc and Analytic Functions

1993. ISBN 3-540-94008-1

Volume 19: **A. A. Gonchar, E. B. Saff** (Eds.)

Progress in Approximation Theory

An International Perspective

1992. ISBN 3-540-97901-8

Volume 18: **W. Hackbusch**

Elliptic Differential Equations

Theory and Numerical Treatment

1992. ISBN 3-540-54822-X

Volume 17: **A. R. Conn, N. I. M. Gould, Ph. R. Toint**

LANCELOT

A Fortran Package for Large-Scale Nonlinear Optimization (Release A)

1992. ISBN 3-540-55470-X

Volume 16: **J. Sokolowski, J. P. Zolesio**

Introduction to Shape Optimization

Shape Sensitivity Analysis

1992. ISBN 3-540-54177-2

Volume 15: **F. Brezzi, M. Fortin**

Mixed and Hybrid Finite Element Methods

1991. ISBN 3-540-97582-9

B3.11.136

Volume 14: **E. Hairer, G. Wanner**

Solving Ordinary Differential Equations II

Stiff and Differential-Algebraic Problems

1991. ISBN 3-540-53775-9

Volume 13: **E. L. Allgower, K. Georg**

Numerical Continuation Methods

An Introduction

1990. ISBN 3-540-12760-7

Volume 12: **C. Brezinski**

History of Continued Fractions and Pade Approximants

1990. ISBN 3-540-15286-5

Volume 11: **J-P. Delahaye**

Sequence Transformations

With an Introduction by C. Brezinski

1988. ISBN 3-540-15283-0

Volume 10: **Yu. Ermoliev, R. J.-B. Wets** (Eds.)

Numerical Techniques for Stochastic Optimization

1988. ISBN 3-540-18677-8

Volume 9: **Z. Ditzian, V. Totik**

Moduli of Smoothness

1987. ISBN 3-540-96536-X

Volume 8: **E. Hairer, S. P. Nørsett, G. Wanner**

Solving Ordinary Differential Equations I

Nonstiff Problems

1993. 2nd revised edition.
ISBN 3-540-56670-8

Volume 7: **D. Braess**

Nonlinear Approximation Theory

1986. ISBN 3-540-13625-8

Volume 6: **F. Robert**

Discrete Iterations

A Metric Study

Translated from the French by J. Rokne
1986. ISBN 3-540-13623-1

Volume 5: **V. Girault, P-A. Raviart**

Finite Element Methods for Navier-Stokes Equations

Theory and Algorithms

1986. ISBN 3-540-15796-4

B3.11.136